I0045300

NOT EVEN TRYING:

THE CORRUPTION OF REAL SCIENCE

NOT EVEN TRYING: THE CORRUPTION OF REAL SCIENCE

Bruce G Charlton

University of Buckingham Press

The University of Buckingham Press

Yeomanry House

Hunter Street

Buckingham MK18 1EG

© The University of Buckingham Press

The moral right of the author has been asserted.

All rights reserved. No part of this publication may be reproduced, stored or introduced into a retrieval system or transmitted in any form or by any means without the prior permission of the publisher nor may be circulated in any form of binding or cover other than the one in which it is published and without a similar condition including this condition being imposed on the subsequent purchaser.

CIP catalogue record for this book is available at the British Library

ISBN 9781908684189

Real science – a definition

Real Science *noun*

Science that operates on the basis of a belief in the reality of truth: that truth is real.

Biographical note

Bruce G Charlton is Visiting Professor of Theoretical Medicine at the University of Buckingham and Reader in Evolutionary Psychiatry at Newcastle University. Bruce has an unusually broad intellectual experience: he graduated with honours from the Newcastle Medical School, took a doctorate at the Medical Research Council Neuroendocrinology group, and did postgraduate training in psychiatry and public health. He has held university lectureships in physiology, anatomy, epidemiology, and psychology; and has a Masters degree in English Literature from Durham. From 2003-10 Professor Charlton solo-edited *Medical Hypotheses*; a monthly international journal that published frequently speculative, sometimes amusing, and often controversial ideas and theories across the whole of medicine and beyond. He has published considerably more than two hundred scientific papers and academic essays in these fields, and contributed journalism to UK national broadsheets and weekly magazines. Bruce G Charlton is author of *Thought Prison: the fundamental nature of political correctness* (University of Buckingham Press, 2011).

Dedication

To the late John Ziman FRS (1925-2005) physicist, and great understander of science; who wrote about Real Science in his book of that name, and was probably the first to distinguish real science from what nowadays calls-itself science but is not.

Note to the reader

This book might strike some people as *bitter* – it is not.

It is however viscerally and unapologetically *angry* – although I hope to have kept this reasonably well under control…

The argument of this book in a single paragraph

Briefly, the argument of this book is that real science is dead, and the main reason is that professional researchers are *not even trying* to seek the truth and speak the truth; and the reason for this is that professional 'scientists' *no longer believe in the truth* - no longer believe that there is an eternal unchanging reality beyond human wishes and organization which they have a duty to seek and proclaim to the best of their (naturally limited) abilities. Hence the vast structures of personnel and resources that constitute modern 'science' are not real science but instead merely a professional research bureaucracy, thus fake or pseudo-science; regulated by peer review (that is, committee opinion) rather than the search-for and service-to reality. Among the consequences are that modern publications in the research literature must be *assumed* to be worthless or misleading and should nearly always be ignored. In practice, this means that nearly all 'science' needs to be demolished (or allowed to collapse) and real science carefully rebuilt outside the professional research structure, from the ground up, by real scientists who regard truth-seeking as an imperative and truthfulness as an iron law.

Contents

Introduction

As a schoolboy and for many years afterwards, I was perhaps as idealistic about science as anyone in recent years – it would not be *much* of an exaggeration to say that I *worshipped* science; since I was an atheist for whom science was the bottom-line description of reality. The great scientists were my heroes – those whom I hoped to emulate.

For me nothing was more *fundamental* than science; everything else was properly to be evaluated from the perspective of science.

Yet now I regard *real* science – the kind of science I used to worship - as a thing of the past; an object of historical study. There are small islands of real science dotted here and there, but with only local and dwindling influence.

To all extents and purposes, I see real science as dead; and what calls itself science as a fake – worse than nothing, because it claims so much: claims indeed the noble mantle of real science.

*

This is not a matter of science having run-out of useful truths to discover. It is that scientists are not any longer trying to discover useful truths.

So, real science has essentially gone. What is now left – a vast, international activity with millions of employed workers and multiple billions of dollars of funding – is so thoroughly corrupt as to be un-reformable.

If enough people care enough about real science to want it back, they will now have to build it *all over again*, from the ground up.

1

*

This book describes the essence of real science: a phenomenon much simpler to describe – yet more difficult to *do* - than you might suspect.

It also charts the course of real science over about a thousand years to its peak in the three centuries up to about 1950, then its extraordinarily rapid – yet dishonestly concealed – collapse down to almost nothing during the past two generations.

It is a remarkable story – covering some of the peaks of accomplishment, and some of the darkest aspects of the human spirit.

Read on...

Understanding science retrospectively

The Owl of Minerva flies only at dusk said Hegel; implying that understanding must be retrospective. Therefore we did not know what science was, nor how it worked (in a philosophical, historical and sociological sense), until real science was already well-advanced towards destruction.

For me, real science is the core of the modern world. Science is the creator and driving force of genuine economic growth (increased efficiency in the production, trade and distribution of essentials), and a significant driver of social change; intellectually science is the crowning glory of modernity; but at the same time and by the same mechanisms, science is responsible for most of the distinctive horrors of the past couple of centuries.

*

My (very basic, to be amplified throughout this book) summary understanding of the rise of real science was that it came from Pagan Greece (epitomized by Aristotle), then through the early Christian theologians - epitomized by the Western Medieval scholastic philosophers (pioneered by Peter Abelard).

It was the Roman Catholic Church that professionalized philosophy as a subject increasingly distinct from theology, and developed the university as institutionally distinct from the monastery (thus dividing education from devotion) – so, the Great Schism (when the Eastern Orthodox and Roman Catholic Churches divided, around 1000 AD) marked the true beginning of modernity.

Then natural science separated from philosophy in the Renaissance era, at around the time of Galileo, and later moved to be focused in Protestant Northern Europe where it first became large, visible and noticeably distinct from about the 17th century.

There were agrarian and industrial revolutions in Britain during the 1700s; and from around 1800 a new world was increasingly apparent: a world characterized by growth in science, technology, the economy, and human capability: the world of modernity. And from this point science became not just a distinct social structure, but a professional career structure.

*

Since the later 19th century, science has, with each generation, broken-up into smaller and smaller specializations, and become more and more career focused.

For a while this specialization led to greater achievement, since it allowed the devotion of more time and effort to solving more manageable problems. Yet each new-generation specialist had been educated in a more generalist tradition – which acted as a drag on the tendency to fragmentation and incoherence.

For a while, therefore, specialization led to greater accomplishment within its individual divisions yet with sufficient integration across these divisions to maintain unity and to check error.

However, specialization continued past this optimal point, and into less-and-less functional fragmentation – such that science lost unity and specialisms lost the ability to serve as mutual checks.

Science gradually became *nothing but* isolated and irrefutable micro-specialisms.

Apparently, therefore, specialization was a *slippery slope* for science: such that once science had stepped-onto the slippery slope of specialization *it could not stop the process*, even when science had slid far beyond the point at which specialization was helpful.

From real science to generic bureaucracy

At some point over the past several decades, science stopped being real and evolved into its current state of being merely a research-based variant of generic bureaucracy.

This was increasingly clear to aware observers from the 1960s, and indeed to the most astute observers (such as Erwin Chargaff) from several decades earlier. But now it is so obvious that only ignorance or dishonesty prevents it being universally acknowledged.

However, bureaucracies are systematically ignorant, and dishonesty is now institutional and compulsory, therefore the disappearance of real science is *not* acknowledged but instead vehemently denied, and steady, incremental progress is claimed!

*

Science presumably always was done among humans – albeit at a very low prevalence; technological breakthroughs have tended to accumulate – albeit with interruptions and local reversals - throughout recorded history; but modernity happened because real scientific breakthroughs came so *thick-and-fast* that increasing efficiency out-ran increasing population – and humanity escaped what Gregory Clark has called the Malthusian Trap.

*

So far, the thesis is relatively uncontroversial. But if modernity depends on the take-off of real science, upon what does the take-off of real science depend?

My answer is *creative genius*.

My understanding is that real science grew fast – especially in the populations of Northern Europe – by recruiting from an increased pool of 'creative geniuses' who were motivated to do science. This I regard as the essential underpinning of modernity.

*

The take-off of science therefore depended on two main things: 1. a sufficient concentration of creative genius focused on scientific problems plus 2. a *modest* degree of cognitive specialization.

That is to say, *smart and creative people working cooperatively on relatively-specific 'scientific problems'*.

And *that*, more or less, is my *definition of science*.

Merely that.

*

So, real science is smart and creative people working cooperatively on scientific problems.

But science proved so useful that it became professionalized, and initially this seemed to accelerate progress considerably. The first few generations of professional scientists from the later 1800s into the twentieth century were immensely productive of significant scientific breakthroughs.

Science seemed very obviously useful – the presumption was that even-more science would be even-more useful...

And so the growth of professional science continued, and continued...

Until it out-grew the supply of creative geniuses and had to recruit from uncreative but very smart people - but continued growing...

Until it then out-grew the supply of uncreative but very smart people, then it had to recruit from uncreative, only moderately smart but hard-working people – but continued growing...

And so on and on, until 'science' consisted of whomsoever who would do specific narrow technical and managerial jobs at the wage and conditions on offer.

That's where we are now...

*

More importantly, professional science initially recruited only those who regarded *the pursuit of truth as an iron law* (and dishonesty was punished by expulsion from science).

Yet, due to professionalization, science increasingly attracted *careerists* rather than truth-seekers.

*

(Truth-seekers are typically resistant to bureaucratic organization; and bureaucratic organization is intrinsically hostile to truth-seekers.)

*

The professionalization of science having eliminated those who were internally-motivated to seek truth; various formal mechanisms and procedures were introduced to try and deal with purely careerist motivations. These mostly amount to peer review mechanisms (peer review = the opinion of a group of senior colleagues).

So, instead of truth-seeking, a filter of *committee evaluations* was applied to ever-more-blatantly-careerist individual behaviour.

And science continued to grow - recruiting less- and less-talented, weaker- and weaker-motivated, less- and - less honest personnel *until...*

... until untalented, unmotivated and dishonest career-orientated professional scientists became *a large majority* within science and included most of *the most successful researchers*; thus careerists took-over the peer review evaluation procedures such as to impose their values; and 'science' became nothing but a 'professional research bureaucracy'.

I wasn't actually doing science

Looking back on 25 years in professional research – I am forced to admit that, although I certainly tried, I wasn't actually doing science.

*

I began professional science in 1984 - or, at least, that's what I thought I was doing.

Since then I worked in and across a variety of fields: neuro-endocrinology (brain transmitters and blood hormones) in relation to psychiatry; the anatomy and physiology of the adrenal gland (especially from 1989), epidemiology (statistics of health and disease, from about 1991); evolutionary psychology (evolutionary aspects of human behaviour including psychiatric illness and the psycho-active drugs, from 1994); systems theory (understanding complex biological organization, from about 2001); and from 2003-10 I edited an international journal of ideas publishing work from the whole of medicine – and sometimes beyond.

*

In all of these areas and some others I found *serious problems* with the existing scientific literature: errors, inconsistencies, wrong framing of problems.

(I don't mean serious problems *in-my-opinion*; I mean that problems objectively, undeniably serious to any honest, informed and competent observer prepared to think for more than five consecutive minutes or two steps of logic – whichever comes first.)

I was not shocked - after all, this is what science is supposedly about, most of the time - providing the negative feedback to correct the wrong stuff.

After all, science is not at any time-point supposed to be wholly-correct, rather it is conceptualized as a system of intrinsic *self-correction*.

(Generating distinctive new lines of true and useful scientific work is what we would all *prefer* to do, in other words to be *original* - but only a few who are both very lucky and very able are able to achieve this.)

*

My assumption was that - as the years rolled by - I would have the satisfaction of seeing the wrong things tested, discredited, discarded and replaced with more-correct knowledge. Error would be eliminated; truth built-upon. So that overall, and in the long term, science would progress.

That is what was supposed to happen.

*

Well, it hasn't happened.

It hasn't happened in *any* of the scientific fields with which I am familiar or of which I have any knowledge. Indeed, instead, much that was true and useful has been lost while much that is utterly worthless – dishonest, incoherent, useless - has thriven.

A few decades ago one could assume that published work was honest and competent (except in specific cases); now one must assume that published work is dishonest and incompetent (except in specific cases).

A few decades ago one could assume that high status ("successful") scientists were honest and competent (except in specific cases); now one must assume that famous and powerful scientists are dishonest and incompetent (except in specific cases).

*

Overall it seems that things have gone *backwards*, and not just slightly.

Yet research activity (personnel, funding, publishing, communicating) have all increased exponentially – doubling in volume every 15 or so years (doubling *every decade* in medical research. And China has *exploded* with research activity in the past 10 years).

So there has been massive expansion of inputs with first stagnation then decline of outputs. Something has gone *terribly* wrong: not just slightly wrong, but terribly wrong.

*

So, I must conclude that although I *believed* I was participating in something called science, something that I thought I understood from the writings of Jacob Bronowski and Karl Popper and from reading the great genius scientists of the past – it turns-out that I wasn't really doing science at all.

I was 'going through the motions' of doing science, true; but the machinery of science was broken, and the work I was trying to do, and the work of those whom I respected, was like a free-spinning-cog – disconnected from mainstream activity.

If real science is that done from truth seeking motives and communicated truthfully, then this kind of science had zero impact on the mainstream.

Get this – real science had become detached from professional research, technology and policy; and (most important) detached from practice: detached from career success, status, funding, publication, prizes and awards...

Real science had become a thing done for subjective personal satisfaction, *merely* a lifestyle choice – nobody else was interested.

*

Maybe real science was being done, maybe it was published, maybe it was cited, maybe it was funded, maybe people made careers from doing it?

But in the end, real science did not make any difference: real science had become just a *private hobby.*

Those few who were lucky enough to find a niche that supported real science did so by accident, not by necessity; and the niches were shrinking all the time.

And we who thought we were participating in the group activity of real science were deluded – pleasantly deluded, perhaps; but deluded.

If not real science, what are professional 'scientists' *really* doing?

The activity of mainstream modern Big Science is most reminiscent of a Soviet Union era organization – such as the grossly unprofitable Polish glass factory I saw on TV being inspected by John Harvey Jones in his TV show *Troubleshooter*.

The factory was producing vast quantities of defective drinking glasses *which nobody wanted*. Nobody wanted to buy them nor even to use them. So the glasses were simply piling-up in gigantic stacks around the factory building – using-up resources, getting in everybody's way, and taking-up all the useful space.

When Harvey Jones was asked what to do, how to make the business profitable, he said the first essential step was *stop making the glasses*.

*

Stop *now*: this very minute, he said. Go out of this office and switch-off the production line, send all the factory workers home (on full pay) for a few weeks, and begin sorting it out.

But so long as the workers were attending daily, beavering-away, filling-in paperwork, with raw materials pouring-in on thundering lorries; the masses of defective glasses were being churned-out, stacked until they were blocking the aisles and preventing anything useful being done... there was no hope.

Better to pay people for doing nothing than this!

*

Same with professional science – *stop it now*, for goodness sake!

Switch-off the assembly line - please!

Stop wasting vast human and physical resources in *piling-up useless stuff that nobody wants*. Better to pay researchers to do *nothing* than this...

*

(Obviously it would be better if people did something useful than nothing – so maybe ex-professional researchers could be dressed warmly and paid to lie across the threshold of closed doors to function as *draught-excluders*?)

*

So, here we have the problem of professional science today – it has been bloated by decades of exponential growth into a bureaucratically-dominated heavy industry Soviet factory, characterized by vastly inefficient mass production of shoddy goods that nobody wants.

And professional science is trundling along, hour by hour, day by day; masses of people going to work, doing things, saying things, writing things, getting funding, spending money, advertising themselves, engaging in petty gossip, intrigues and back-stabbing...

Science is hopelessly and utterly un-reformable while it continues to be so big, continues to grow-and-grow, and continues uselessly to churn out ever-more of its sub-standard and unwanted goods.

Not even trying

Switch it off: stop making the defective glasses: *now...*

The pervasive dishonesty of modern 'science'

How did we get from useful and real science to useless research bureaucracies generating hype and spin for the public relations industry?

Anyone who has been a scientist for more than 20 years will realize that there has been a progressive, significant and indeed *qualitative* decline in the honesty of communications between scientists, between scientists and their employing institutions, and between scientists and their institutions and the outside world.

In a nutshell – science has gone from being basically honest to basically dishonest (and in the process gone from being real science to professional research).

*

Naturally enough, the pervasive atmosphere of dishonesty has long since led to scientists being dishonest *with themselves* - and once this happened the situation of endemic corruption itself became wholly deniable.

(The primary and fundamental act of scientific dishonesty is: *denial of the pervasive reality of scientific dishonesty.*)

The situation now is that what calls itself scientific research is *essentially* dishonest, not incidentally so; such that honest (real) science is on the one hand very rare and on the other hand it has negligible impact on the conduct of mainstream research.

(From my experience it seems that real science is nowadays more likely to be actively-suppressed, and real scientists systematically persecuted, than for either to be influential.)

*

More exactly, mainstream research is not so much dishonest as *non*-honest: it is simply *unconcerned* by matters such as seeking truth and rigid truthfulness in its discourse. Mainstream research is not *about* truth – it is doing other things.

*

So *of course* modern 'science' is dishonest – why on earth should it be honest when it is *not even trying* to be honest? Research is not being done to find the truth, experiments are not done to test the truth, scientific ideas and results are not written-up in order to communicate the truth.

Truth doesn't go into it. Why on earth, then, should anybody imagine that truth will come out of it?

Mainstream professional research is no more about honesty than advertising, politics or official statistics are about honesty.

Which is to say that in modern professional research there is *just enough narrowly-factual accuracy* to render deniable its basic and motivated dishonesty.

*

Yet real science *must* be an arena where truth is the rule; or else the activity simply stops being science and becomes something else – professional research, a job, a bureaucratic institution, an arbitrary activity done to justify funding.

18

The honesty of *real* science is not merely a desirable feature, an optional extra, it is intrinsic to the activity – real science is built-around honesty as its core ethic.

Discard honesty and there is nothing left of science.

*

Mathematicians know that even random (accidental, undirected) errors multiply very rapidly – third and fourth generation analogue recordings are more crackle than music and beyond that there is just crackle. Yet random errors occur equally in both directions from the signal, noise tends to cancel out; and so can be averaged. The signal can be retrieved from even a great deal of noise.

But *systematic* error cannot be averaged, because there is no reason to assume that falsity is equally distributed on either side of the truth: and dishonesty is a systematic error.

Truth cannot be retrieved from a mass of lies – so without strict honesty, truth is simply *lost...*

Communications within the science profession

The most egregious domain of untruthfulness is probably where scientists speak or write about their own work.

When modern researchers are preparing applications for funding, there is clearly no notion that they should be trying to communicate the truth. The idea would be regarded as ridiculous! The whole motive and rationale of the exercise is to write a *successful* application: in other words to get yourself money by selling (what you claim are) your research results and plans.

The *veracity* of what is being claimed is merely *a means to an end*. The funder neither expects truth nor does the applicant expect to write truth – *grantsmanship* is thus a kind of game (albeit with high stakes) where one side sets up the rules, and the other side tries to be as dishonest as it can get away with, while sticking to the letter of the rules - then the first side tries to catch them out in an inconsistency.

Modern research grant proposals therefore resemble the official accounts of organized crime – everyone knows that they are intentionally and carefully faked, but the auditors are allowed only to check for internal consistency among the lies. *Consistent* lying is fine – indeed admired and rewarded.

So long as the information in grant proposals and research publications has been thoroughly *laundered*, then everybody is happy (well, 'everybody' who has influence over career success – and for modern researchers that is everybody-who-matters...).

*

When a modern researcher describes the nature and significance of his research to another researcher in the same field, for example in a writing a publication or speaking at a conference or in the casual interactions of scientific life; or to a bureaucrat or official who might directly or indirectly influence his pay and conditions (say a university administrator); or to a journalist or media person, or even to a random member of the public – there is no notion of the modern researcher trying to be truthful about the nature and significance of his research.

Trying *strictly* to be truthful would indeed be regarded as evidence of naiveté, or – if persisted-with - actively dangerous.

Modern research communication is *strategic* – it is a *means to an end*: and the degree of veracity of what is being said is controlled by the requirements of that end.

Dishonesty as pervasive, endemic

Once they have been observed, selected and trained; *real* scientists are unreflectingly honest and will trust each other; but their honesty is also enforced by a multitude of informal processes – if, after being trusted, that trust is *betrayed* there is the permanent sanction of exclusion from real science. From that point, real scientists will simply ignore you.

*

In discussing the dishonesty of modern 'science' it is tempting to focus on cases of 'fraud' – especially instances in which specific researchers have fabricated, invented or deliberately distorted their results for personal gain.

It is tempting but misleading, because it assumes honesty as a baseline. While real scientists are indeed habitually truthful, modern professional researchers by contrast are *not even trying* to be truthful.

Truth is a positive value. However, at most, modern researchers are trying *not* to be factually *incorrect* – which is as different from trying to be truthful as a scandal-mongering tabloid 'investigative' journalist is different from Einstein.

This is not a subtle matter. Nor is it a matter for debate. It is absolutely plain and obvious on a day-by-day level in the *conduct and conversation* of modern researchers. Compared with real scientists, the mass of modern researchers (including, perhaps especially, scientific leaders) are neither motivated nor regulated by truth, nor do they speak about truth, nor do they discriminate on the basis of truthfulness.

*

You doubt this?

Just watch! Just listen! Just read! So long as you can tell the difference between on the one hand someone trying to be as truthful as they can be, and on the other hand someone trying to *sell something* – then it is a no-brainer.

Dishonesty with oneself

So pervasive are the petty misrepresentations and cautious lies, it is evidence that many scientists are now dishonest even with themselves, in the privacy of their own thoughts.

Such things can happen to initially honest people either by force of habit, or because they never knew any better (never having *met*, leave-aside *worked-with*, a real scientist); and because lies breed lies in order to explain the discrepancies between predictions and observations, between claims and outcomes.

*

Lying to oneself may be one cause of the remarkable incoherence of so much modern scientific thinking, when coherence is evaluated across the whole range of human knowledge.

(The coherence of modern science is restricted to the micro-specialty; where it is the artificial result of laundering rather than natural consequence of honestly reporting perceived reality.)

It is much easier to be coherent, and to recognize incoherence, when discourse is uncontaminated by deliberate misrepresentations. There is less to cover-up.

Most people can think-straight only by being completely honest with themselves and with everybody else. Maybe straight thinking doesn't matter in some areas of life – but science is about straight thinking or it is nothing.

If scientists are not honest even with themselves, then their work will be a mess – or rather, because modern researchers are *not* honest with themselves their work *is* a mess.

*

Scientists are usually too cautious and timid to risk telling outright *lies* about important things, or to invent and emphasize fake *data*; but instead they push the envelope of exaggeration, selectivity and distortion as far as possible. And tolerance for this kind of untruthfulness has greatly increased over recent years.

So it is now routine, normal, indeed *required* behaviour for scientists deliberately to exaggerate, to 'hype' the significance of their status and performance, and 'spin' the importance of their research.

The envelope of exaggeration is now extended to the not-impossible: so if it is in reality *not-impossible* that my research *might* (under highly implausible but *not-impossible* combinations of conditions) assist in some way in curing cancer... then it is nowadays permissible (in a 'good' cause – i.e. when it is expedient) to present the research as being *progress towards* curing cancer.

In sum, when a modern researcher says 'my research is progress towards curing cancer' it *really* means 'it is not impossible that my research could conceivably count as progress towards curing cancer'.

*

Furthermore, it is entirely normal and unremarkable for ordinary 'scientists' to spend their entire professional life doing work they know in their hearts to be trivial or bogus – preferring that which promotes their career over that which has the best chance of advancing science.

Indeed, it is entirely normal and unremarkable for the *best* modern 'scientists' to spend their entire professional life doing sub-optimal work they know in their hearts to be less scientifically ambitious than they are capable of.

In a nutshell the most successful modern researchers have replaced scientific ambition with career ambition.

Far from being frowned-upon, such gross and treacherous misapplication of research effort is positively *encouraged*, nay enforced, and not just sometimes but as the norm in many places and by many people, including what are supposed to be the best places for research (universities and other institutions); because careerism is a more reliable route to high productivity than real science.

In fact it may be impossible to get a job, or get tenure, or promotion - except by dumping idealism and scientific ambition and embracing low-risk careerism.

*

Indeed, senior scientists *in the best places* are clever, hard-working and intelligent enough rapidly to become expert at hyping mundane research to create a misleading impression of revolutionary importance. Far from resisting, or fighting, the degradation of science; the senior researchers at the 'best' places have *led* (indeed *driven*) their subordinates into a morass of corruption, like so many demonically-possessed Gadarene swine.

It is a kind of Gresham's Law at work; when dishonest research is treated as if it were real science; then bad research drives out the good.

*

So, in real science there is, there *must be*, zero-tolerance for dishonesty and zero-compromise with truthfulness.

Truth-telling and truth-seeking must not be regarded as mere ideals within science, but as *iron laws*, continually and universally operative.

Causes of dishonesty in science

Although some scientists are selfishly dishonest simply in order to promote their own careers, for most people quasi-altruistic arguments for lying (dishonesty in a good cause of helping others, or to be an agreeable colleague) are likely to be a more powerful inducement to routine untruthfulness than is the gaining of personal advantage.

For example, scientists are strongly pressured to be less-than-wholly-truthful for the benefit of their colleagues or institutions, or for official/political reasons – for example in fund-raising, or complying with inspections or external research evaluations.

(And in areas of science that impinge on the taboos of political correctness or the imperatives of 'progressive' politics, honesty is punishable with extreme disincentives – career termination, media-orchestrated vilification, legal prosecution, threatened and actual violence.)

*

Often, scientists are unable (without attracting severe sanctions) to opt-out of administrative or managerial exercises which all-but *insist*-upon dishonest responses – and for which colleagues *expect* dishonesty in order to promote the interests of the group.

Failure to comply would be seen as selfish scrupulousness at the expense of colleagues. There would be no support from scientific leaders – whose careers stand to benefit most from success in administrative or managerial exercises.

Project leaders may feel responsible for raising money to support the livelihood of their junior team members; and feel obliged to do whatever type of research is most generously funded, and to say or write whatever is necessary to obtain that funding.

Failure to do *whatever it takes* to secure funding or survival in a bureaucratic system would be seen as a failure to provide for your dependents – as sacrificing peoples' livelihoods on the altar of your own *smug* desire to feel virtuous...

*

So, in a bureaucratic context where cautious and consistent dishonesty is rewarded, strict truthfulness is *taboo* and will cause trouble for colleagues, for teams, for whole institutions.

Because when *everyone else is exaggerating* their achievement then any precisely *accurate* person will be judged as even worse than their already modest claims.

If every fourth rate scientist is *claiming* to be third rate – but after inflation-adjustment is *judged* to be fourth rate; then honestly to label oneself as fourth rate would lead to being to be judged as fifth rate - on the assumption that you, like everyone else, must be indulging in hype.

In this kind of situation, individual truthfulness will be interpreted either as simply stupid, or as an irresponsible indulgence.

*

Clearly then, even in the absence of the sort of direct coercion which prevails in many un-free societies, scientists may be subjected to such pressure that they are more-or-less *forced* to be dishonest; and this situation can (in decent people) lead to feelings of regret, or to shame and remorse.

The only alternative is some species of martyrdom.

*

This is a situation which leads decent people to feel shame and remorse.

Unfortunately, shame may not lead to remorse but instead to *rationalization*, to self-exculpation, to the elaborate construction of excuses - and eventually a *denial* of dishonesty. In other words, shame may lead to aggressive hypocrisy.

But eventually the situation leads many to cynicism; hypocrisy is abandoned as ludicrously implausible – and there is a cynical *advocacy* of dishonesty. Such cynics feel they are merely being honest in advocating open dishonesty, because everyone is doing this anyway. Better – they think – to be a cynic advocating evil than a hypocrite pretending to good.

Yet, whatever are the motivations and reasons for research dishonesty, it has been by such means that modern 'scientists' have become inculcated into *habitual* falsity; until people have become used-to dishonesty, don't notice dishonesty, eventually come to expect and finally insist-upon dishonesty.

Roots of dishonesty in science – the role of peer review

My belief is that science has rotted *from the head down* – from the top to the bottom - and therefore blame mostly lies with senior 'scientists'.

The careerism of senior 'scientists', and their abandonment of the Iron Law of truthfulness, has been the main cause of the now pervasive corruption of science (not least because the senior appoint the junior, the bosses choose the minions).

So the roots of dishonesty in science constitute a 'treason of the clerks' phenomenon.

*

While the ultimate cause of the treason has been the abandonment of truth conceived as a *transcendental value* – as I argue below – the proximate mechanism by which corruption has been implemented was peer review.

Since the middle twentieth century there has been massive expansion in the size and influence of peer review, peer review infiltrated into all the major scientific evaluations – peer review has become the self-perceived core process of science.

Yet peer review is no more, no less, than the *opinion* of senior scientists. And not individual judgment, but a procedure for gathering opinions of a *group*, followed by some kind of more-or-less formal, more-or-less explicit procedure for deriving a single decision from the group of opinions: by vote, by veto, by some kind of weighted quantification, by an impressionistic judgment of the decision, or whatever.

In practice, most peer review is a 'black box' mechanism – and all the more effective for its unknown operations. A question is fed-into the black box of peer review, some senior scientists deliberate in some way and some answer emerges – an answer that is impossible to critique yet regarded as authoritative (as if a *committee* of senior scientists constituted a kind of super-multi-brain with magically-combined wisdom and expertise!)

The essence of peer review is therefore the 'peers' – which implicitly means a plurality of senior figures from (broadly) the same domain or field of research endeavour; and the 'review' element which in some way derives a bimodal or categorical evaluation from the plurality of opinions.

*

To put it another way, the triumph of peer review is a triumph of the committee over the individual, of procedure over judgment, of the selective and explicit over the unbounded and implicit.

The even-more-significant aspect of peer review is the rhetorical success of implying that a committee procedure is more objective and more *valid* than individual judgment; the almost-wholly successful trick of disguising that peer review is pure opinion, and therefore just as 'unreliable' and prone to corruption as individual judgment – but that in fact peer review is worse than individual judgment for the same reason that a committee decision is intrinsically worse than an individual decision: because the committee decision is removed from individual responsibility, hence removed from responsibility altogether.

(Responsibility is an attribute of individual authority. Without I.A. there is no responsibility – merely a legal contract.)

Yet peer review is *neither necessary nor sufficient* as a definition of science, it is *orthogonal* to science; and *therefore* domination by peer review marks the disappearance of 'real science' and the inclusion of its activities within the system of large, complex trans-national bureaucracies.

*

So peer review does not *solve* the problem of subjectivity; rather it *replaces* potentially responsible individual subjectivity with necessarily irresponsible group subjectivity.

Thus the advantage of peer review is precisely the opposite of its propaganda – peer review has become universal *because* it is irresponsible, not despite this.

For peer review; irresponsibility is a feature, not a bug.

*

Overall, senior scientists have set a bad example of untruthfulness, self-seeking and *lack of principle* in their own behaviour, and (surely not unrelated) they have also tended to administer science in such a way as to reward hype and careful-dishonesty, and punish modesty and strict truth-telling.

Some senior scientists have laudably refused to compromise their honesty, however they have done this largely by quietly 'opting-out', and not much by using their power and influence to create and advertise alternative processes and systems in which honest scientists might work. They have not *exposed* the pervasive and mandatory dishonesty of modern 'science.

Presumably they *began* by not wanting to discredit what real science still remained, but ended by colluding in the disguise of the non-scientific nature of pseudo-scientific professional research.

But, to be fair to the honest real scientists, those that did speak out loudly and clearly – such as Erwin Chargaff - were first marginalized, then ridiculed, then completely ignored and forgotten – as being embittered failures, motivated by 'sour grapes' and envy...

*

Peer review - of ever greater complexity, hence irresponsibility - has now been applied everywhere: to academic education and research training, job appointments and promotions, to scientific publications and conferences, to ethical review, to research funding, to the allocation of medals, prizes and awards.

And peer review processes are set-up and manned by senior scientists. In a sense, peer review (where it matters, where it makes a difference to policy and practice) simply *is* monopolization of all evaluation, reward or punishment processes by senior scientists; yet not as autonomous individuals but as components of a *process* which nobody-in-particular controls.

This seems something like the worst of all possible worlds; most of the actual disadvantages of tyranny but without any of the potential advantages of having 'somebody' in control.

Modern 'science' is *de facto* dishonest

Of course not every single modern scientist is dishonest, and not every last branch of professional science is corrupt.

However, in practice, *they might as well be.*

By 'in practice' I mean to make that distinction that, from the transcendental and ultimate perspective, corruption is an evil and thus every individual scienctist who holds-out against the prevailing dishonesty *counts.*

Yet when honest scientists and truthful specialties are *disarticulated* from the processes of mainstream science (especially from the outputs of peer review processes) then these do not affect functioning of the *system.*

The existence of a few honest souls does not refute the charge of general scientific dishonesty – just as the existence of a handful of impartial judges does not refute the charge of systemic legal corruption.

(After all the rare honest judge can be, often is, over-ruled by corrupt superiors. It happens.)

Peer review is neither a necessary nor sufficient part of real science

I have often read comments which state explicitly, or assume implicitly, that peer review is what sets science apart from other (less valid) modes of knowledge.

Yet this is simply, observably, demonstrably false. Peer review is neither necessary nor sufficient to real science.

Peer review is *not necessary*, nor was peer review a feature of science in its golden age, when science worked best – most effectively and efficiently.

Old writings never mention anything like modern peer review. In those eras decision making was mostly, sometimes wholly, *individual and personal* (with certain exceptions where a 'collegial' method of decision making was used to allocate goods that were generated and controlled by an institution).

And peer review is *not distinctive* to science, but is indeed (very obviously – I would have thought) found in *all* academic subjects nowadays; and is characteristic of many formal bureaucracies. Indeed, peer review is perhaps the defining feature, the hallmark of modern bureaucracies in which personal responsibility has been replaced by (deliberately, not accidentally) unaccountable committee procedures.

*

The over-expansion and domination of peer review in science is therefore a sign of scientific decline and decadence, not (as so commonly asserted) a sign of increased rigour.

Peer review as the ultimate arbiter represents the conversion of real science to generic bureaucracy; a *replacement of testing knowledge by opinions about knowledge*; a replacement of objectivity by subjectivity – imposing a procedural but arbitrary subjectivity rather than having individual subjectivity linked to responsibility.

And the increased role for *de facto* irresponsibility in science has created space into which dishonesty has expanded.

When modern 'science' is not honest, as it typically is not, then peer review ensures that *nobody-in-particular is identifiably to-blame* for the situation.

*

As well as from the careerism of senior scientists, inducements to dishonesty have also come from outside of science – from politics, government administration and the media (for example), all of whom are continually attempting to distort science to their own agenda and covert real science to the service of their power.

At present, the situation in the UK is that a researcher cannot get money from a government source without perjuring themselves.

(Naturally, I refer to perjury by appropriately scientific criteria of departure from absolute *truth-full-ness*; and not by irrelevant legal criteria of perjury as provable-lying.)

(The skill of scientific perjury, as practiced by the most modern successful researchers, is indeed precisely to commit scientific perjury while avoiding legal perjury.)

*

But whatever the origin of the pressures to corrupt science, it is obvious that the scientific leadership have themselves been corrupted and co-opted.

The alternative would have been inflexible resistance on a matter of principle – the principle of truth-seeking and truth-speaking as an Iron Law intrinsic to science, even to the point of 'martyrdom'.

Notable individuals from past generations of scientists did indeed stand up for their beliefs to the extent of being sacked, imprisoned, exiled or even killed. We moderns can only stand in awe of such principled behaviour.

But modern 'scientists' have been kept in-line without any need for recourse to such draconian measures. The mildest of implied threats have been enough to convert real scientists into careerist drones.

In real science truth must be a transcendental value

Why, how did past generations of real scientist behave *so much better* – so much more truthfully - than modern professional researchers?

I have come to believe that real science depends for its long-term success on an explicit and pervasive pursuit of the ideal of transcendental truth.

'Transcendental' implies that a value is outside the material world; is real, stable and ultimate – it is aimed-at but can only imperfectly and imprecisely be known, achieved or measured.

So, transcendental truth is an ideal but actual thing, located outside of science; beyond and above scientific methods, processes and peer consensus.

*

Transcendental truth is not, therefore, evaluated *by* science; but is instead the proper *aim* of real science. It is *regulatory* of real science.

(Technically, transcendental truth is a metaphysical assumption. And that-there-is-no-such-thing-as-transcendental-truth is also a metaphysical assumption.)

Especially truth is the proper aim of scientists as individuals. In other words, science should be a social system constituted by individual scientists who are dedicated truth-seekers: whose practice of science includes 'truth talk' that references current actuality to ideal aspirations and who practice 'the habit of truth'.

Real science is not, therefore, made of institutions, nor organizations, nor of rules, methods nor processes – real science is made by, done by, individuals: people working together to discover and communicate reality.

Jacob Bronowski on the habit of truth

Jacob Bronowski (1908-1974) invented the term 'the habit of truth' to describe the fundamental and distinctive ethic of science: the main foundation upon which was built the success of science, providing the means (knowledge) for mankind to shape the natural world.

Bronowski emphasized this, since it was (and is) often imagined that science is a morally *neutral* activity. This is wrong. Because, although scientific knowledge is indeed morally neutral (and can be used for good or evil), the *practice* of science (including being a scientist) is certainly a moral activity - based on the habit of truth.

He argued that for science to be truthful as a whole it is not sufficient to aim at truth as an ultimate outcome, scientists must also be habitually truthful in the 'minute particulars' of their scientific lives.

The end does *not* justify the means, instead Bronowski argued that the means are indivisible from the end: scientific work is 'of a piece, in the large and in detail; so that if we silence one scruple about our means, we infect ourselves and our ends together'.

*

Bronowski's insight was that – to be successful in terms of the test of shaping the natural world – each and every scientist in his scientific communications must speak the truth as he understands it.

To put it another way – scientists must be *trying* to seek the truth, *trying* to be truthful – all the time and about everything.

Indeed, I think it likely that the social structure of science is at root *nothing more* than a group of people investigating reality who seek truth and speak truth habitually (and if, or when, they cannot be truthful, they will either state this or say nothing).

*

Bronowski perceived that societies which abandoned, indeed persecuted, the habit of truth – such as, in his time, the USSR and Nazi Germany – paid the price in terms of losing their ability to perceive or generate the underlying knowledge of reality which forms the basis of shaping the natural world.

(Note – these were societies which had had the habit of truth in science at one time, but then 'lost' it; or rather – like ourselves – actively *crushed* it.)

This declining ability to shape the natural world was concealed with propaganda, but such concealment could only be temporary since the cause of the decline was strengthened by every attempt to deny it.

But, the scientific failures of Germany, and especially the USSR, were obvious in comparison with the USA and the UK – however, when the USA and the UK abandoned truth (along with pretty much all other places) then there was no comparator; the effect was not obvious, could more easily be hidden...

(After all, *somebody* will be awarded a Nobel Prize every year – whether or not anybody *deserves it*, whether or not there is any *real* science being done in the field in question.)

*

Having grown up strongly under the influence of Bronowski (for good and for ill) and also this distinctive morality of science, I have witnessed at first hand the rapid loss of the habit of truth from science.

At first I saw an *encapsulated* loss whereby scientists continued to be truthful with each other (that is, truthful in the sense of speaking the truth as they see it) while lying to outsiders (especially in order to get grants, promote their research, and to deflect criticism)...

Then scientists stopped being truthful with other scientists (who were now seen as competitors, gatekeepers, potential patrons)...

After which the situation degenerating swiftly to the final surrender whereby scientists are no longer truthful *even with themselves*.

*

At the same time I have seen hype (i.e. propaganda) expand from being merely a superficial sheen added to real science in order to make it more interesting to the general public, to the present situation where hype *defines* reality for scientists (as well as everyone else) – where propaganda is so pervasive that nobody can know what – maybe nothing at all, or the opposite to the propaganda – lies beneath it.

There is, indeed, no 'beneath' since by now hype goes all the way through science: from top to bottom, inside and out.

A 50 year *experiment* in excluding transcendental truth from scientific discourse

Although the ultimate scientific authority of a transcendental value of truth (located outside of current scientific practice) was a view almost universally held by the greatest scientists throughout recorded history, and was a frequent topic of discourse among scientists and in the literature until the mid-20th century; modern science has pretty much dispensed with the idea of truth.

References to truth in an ultimate sense have by now been all-but banished from professional scientific literature and discourse; being regarded by a younger generation of hard-nosed and technically-orientated researchers as wishful, mystical and embarrassing at best – and hypocritical or manipulative at worst.

Instead, all disputes are constrained to operate within an evaluation system of proximate methodology and peer approved standard practice.

*

Such exclusion of references to truth from scientific discourse could be regarded as an *experiment* which has been gathering support for about 50 years – although the overlapping of scientific generations meant that senior scientists continued to discuss truth in a transcendental fashion at least into the 1980s, and a handful still continue.

The experiment in exclusion of truth talk was driven (presumably) partly by the desire for *greater efficiency* (the desire for less metaphysical chit-chat and more *hard science*) – and partly on the belief that transcendental values *serve no practical function* – merely waste time and energy, confuse and mislead. The assumption was that science could more-efficiently be done using just internal, professional (within-science) evaluations.

Partly it was also driven by the increasing prevalence of materialist atheism – such that 'scientists' no longer believe in transcendental reality; indeed some modern 'scientists' seem *not* to believe that there is any reality separate from social structures that describe and define what-counts-as-truth. They seem to operate on the basis that reality is 'socially constructed'.

Modern 'scientists' are not interested in whether something really is true; they are interested *only* in whether peer review *says* it is true – they are interested only in whether something is fashionable, funded, publishable in high-impact fora, and likely to attract jobs, promotions and prizes.

Even those who publicly oppose and ridicule the idea of social construction of 'reality' *behave* as if a vote from a peer review committee of senior 'scientists' is the nearest possible approximation to truth – which is a view as close to pure reality-denying nihilism as makes no difference...

*

This profound shift within science was described tellingly in *Real Science* by the late John Ziman (1925-2005) (from whom I took the sub-title of this book). Ziman was a British physicist of great distinction as well as a philosopher and sociologist of science, and on the advisory board of *Medical Hypotheses* when I was editor.

Ziman termed the transformation in science during his lifetime a change from 'academic science' to 'post-academic science'.

Academic science is what I call 'real' science; post-academic science is what I call 'professional research'.

*

In Ziman's description, *post*-academic' discourse is implicitly framed such that questions of truth have lost their meaning. It is a type of Big Science – focused on the organization and funding of projects.

Real Science memorably describes the transformation in the fine texture of a successful scientist's life, the day to day *activities*.

The *old style* 'academic' or real scientist does science – tries to discover, theorise and describe the truth about reality.

But the typical day of a *modern*, professional-researching post-academic 'scientist' is non-overlappingly distinct from this. It is, in essence, the life of a bureaucrat, of a manager – combining personnel administration and project organizing with public relations, arranging for publication, fund-raising, publicity and presentations.

*

The lack of any anchor from research practice to transcendental truth has rendered many areas of modern 'science' a kind of 'glass bead game' (to use the term from Herman Hesse's novel), comprising research disciplines that are free-spinning cogs with little or no explanatory, predictive or manipulative connection with the natural world.

By its ultimate reliance on professional evaluators (various different versions of peer review applied to research funding, publication, prizes, promotions, etc.) modern 'science' has become *structurally indistinguishable* from academic literary criticism:

both being arcane, technically non-intuitive and rigorous, sometimes intellectually brilliant – but ultimately internally-validated *fashion-driven high brow pastimes* comprised of ringing variations for the sake of career advancement.

The experiment in trying to do science without reference to transcendental truth has therefore failed utterly. In discarding transcendental truth, science discarded what had made it science.

What is left over is a *fundamentally dishonest sham* which tries to claim the distinctive validity of real science without submitting to the iron discipline of truth.

But is truth really *true,* or was it just a convenient fiction?

It seems that transcendental truth is needed in science, for science to work, for science to remain science.

In *Human Accomplishment* (2003), his analysis of the most influential figures in the Arts and Sciences, Charles Murray makes a strong case that belief in the reality of transcendental values is characteristic of those historical eras which are characterised by the highest level of achievement.

Only when science is truth-seeking can its practice mobilize the most profound dedication from its practitioners – a level of motivation far greater than that elicited by peer-approval-seeking science, or science done from a familial or social sense of duty.

Recall; when scientists believed in truth they would historically suffer hardship (sometimes extreme hardship, prison, even death) for their scientific beliefs. But nowadays even the mere possibility of being passed over for a grant or promotion is sufficient to terrify 'scientists' into submission.

Another reason for valuing truth is the need for science as a social system to tolerate (and if possible actively support) individuals who seek truth – even when this generates greater risk and a short term reduction in performance.

Likewise the discipline of transcendental truth enables science to tolerate the fact that many brilliant and creative scientists will often have unworldly, erratic or abrasive personalities.

In other words, only the *living presence of truth* in the daily practice of science may provide a higher context for decision-making in which considerations of social expediency can potentially be transcended.

*

But despite these advantages, the 'big question' for any modern scientist is whether transcendental truth really is 'true' or is merely a convenient fiction.

By 'convenient fiction' I mean the idea that *even if* it could convincingly be argued that scientists work better when they believe in transcendental truth; such 'truth' is *actually* no more than a delusion, albeit a useful delusion.

The convenient fiction argument is that in reality there is no such thing as truth but it is a good thing for science and for society when scientists act *as if* truth is real.

*

The discussion then moves beyond science, and to the presuppositions of science; moves to a level of the basic understanding of things – in other words, to metaphysics.

Early scientists generally assumed (I mean they assumed at a *metaphysical* level – as their conception of the nature of reality) that the truth was reality - a property of the universe created by a god.

Truth (knowledge of reality) was communicated in outline to humans partly by being in-built (by god) as human nature and partly from divine revelation; truth was understood by means of reason (which was valid because also god-given), and applied to the study of Nature by god-given human ingenuity.

Early scientists therefore believed in both god/s and truth.

Later scientists (from the late 19th century into the early 20th century) were atheists about god but realists about truth. For example Albert Einstein had an abstract, or pantheistic view of an ordered universe and a belief in the fortunate (but not god-given) rational and intuitive ability of humans to understand the nature of reality.

*

Most of these scientists of several generations ago were *theists*, more or less; believing in an impersonal god who created order (and the order was real – for example mathematics or the laws of physics were real); but setting-aside divine aspects of individual salvation, meaning and purpose.

Another generation or two onwards, and most of the best scientists were atheists about god and also did not believe in the reality of truth. They disbelieved in both God and truth, nonetheless the best scientists continued to *behave* as if they did regard truth as real. For example Richard Feynman was not religious and seemingly did not believe in *transcendental* truth but anyway lived and worked by a strict *personal* ethic of truthfulness and truth-seeking.

Modern scientists have abandoned all this as so much useless baggage. They are atheists about god, relativists about truth, and careerists in their behaviour: they neither believe, nor behave as if they believe, in transcendental truth.

*

So, the historical sequence was: theism, deism, atheism.

Deism perhaps enabled the greatest science; but deism was temporary and *en route* between theism and atheism.

How a scientist behaves is clearly more important than his or her belief system. Einstein and Feynman behaved (with respect to science) in an exemplary fashion.

Yet viewed through the 'retrospectoscope' I am not convinced of the coherence or long-term sustainability of Feynman's views – nor even Einstein's.

To be truthful yet believing neither in transcendental truth nor in a personal relationship with deity, now just looks like another unstable *phase* between theism and atheism – a perspective restricted to the transitional generation of people who were *brought-up religiously then abandoned it in adulthood.*

The next generation, their children and grand-children – born from the mid-twentieth century onwards, were brought-up as secular materialists, have moved decisively to atheism *and also to non-truthfulness.*

*

In a nutshell, it seems that there are several ways to live by transcendental truth – ranging from formal religion to a pragmatic assumption that it is expedient to act as-if truth were real.

But some belief systems relating to truth are more stable and coherent than others, and some belief systems are more powerfully motivating than others.

For scientists, the crucial matter is that each real scientist must, must, *must* (for whatever reason) work according to a binding personal ethic of the importance and reality of transcendental truth – that truth lies beyond and above science; and science must be practiced according to this reality

Not even trying...

While wanting to know the truth does not mean that you *will* find the truth; on the other hand, if scientists are *not even trying* to discover the truth - then the truth certainly will *not* be discovered.

Even if stumbled-upon, or tripped-over by happy accident by someone *not looking for it*, then truth will *fail to be recognized* as true.

If the truth is in a particular direction, then there are many more directions (an infinite number of them) where the truth *cannot* be found; so when a researcher is *not* looking for the truth, the chances of finding it are one to infinity.

In a nutshell, there are an infinite number of ways to be wrong, but only one way to be right.

(Of course, nobody is ever completely right – but even to be *approximately* right entails the objective reality of universal and eternal truth.)

*

'Truth' can be defined as 'underlying reality'. Science is not the only way of discovering truth (for example, philosophy is also about discovering truth - science being in its origin a sub-specialty of philosophy) - but unless an activity is trying to discover underlying reality, then certainly it cannot be science.

But what *motivates* someone to want to discover the truth about something?

The great scientists are all very strongly motivated to 'want to know' about reality, and this drove them to great efforts, risk, hardship - and kept them at their task for decades.

Why scientists should be interested in one thing rather than another thing remains a mystery – but what is clear is that this interest cannot be dictated but arises from within – having arisen it can be encouraged but not re-directed.

Real science is a vocation.

*

Francis Crick commented that you should research that about which you *gossip*, James Watson commented that you should avoid subjects which *bore* you.

Their point was that science is *so difficult*, that when motivation is deficient then problems *will not get solved.*

You need, *you must have*, spontaneous positive interest (the gossip test) and you cannot solve problems that bore you because real science is too hard to succeed without the benefit of spontaneous interest.

Motivation needs all the help it can get – hence *real science cannot be dictated.* It cannot be planned.

The directed provision of research funding and the implementation of research *strategy* can certainly make people 'do research' in a particular field; but it cannot make them do real science.

*

But there is an opposite assumption at work in mainstream modern 'science'; the idea that professional researchers should *properly* be motivated by *career incentives* such as appointments, pay and promotion – and *not* by their intrinsic interest in a problem.

This is rationalized on the basis that personal motivations are a probable source of *bias*.

Well, maybe they are – but without personal motivation you don't get science at all.

The way to get valid science is not to employ people who care so little about what they study that they are *impartially uninterested in everything* and will believe and work on anything.

The way to get valid science is to have a group of inevitably-biased people working together to seek the truth – *the motivation to seek truth will find ways to deal with bias*, as was seen many times in the history of science.

In reality, the bureaucrats who run science just *do not want* vocationally motivated scientists as employees, since they are intrinsically awkward individuals - precisely for the reason that their beliefs and activities can neither easily nor wholly be shaped by career incentives.

*

By contrast, careerist research drones who want to be 'successful' will *do whatever they are told to do* and will not do what they are punished for doing. Careerist research drones would not, for example, insist on trying to discover the structure of DNA when they were supposed to be doing other things – as did Crick and Watson.

Modern pseudo-scientific bureaucrats would try very hard *not* to employ anyone with the awkward personality traits of Crick and Watson, and indeed very few modern researchers are of that type.

Thus everything runs smoothly, people do exactly what they are supposed to do – and the only problem is that *zero real science gets done...*

However, that 'perception' is easily fixed by public relations, hype and spin.

The peer review cartel

The modern scientist is supposed to be a docile and obedient bureaucrat and is trained and selected for that purpose – cheerfully switching 'interests' and tasks as required by the changing (or unchanging) imperatives of funding, the fashions of research and the orders of his master.

What determines a modern scientist's choice of problem? Essentially it is *peer review* – the modern scientist is supposed to do whatever work that the cartel of peer-review-dominating scientists decide he should do and reward him for doing.

This will almost certainly involve working as a team member for one or more of the peer review cartel scientists (or their out-sourced 'suppliers'); doing some kind of allocated micro-specialized task of no meaning and zero intrinsic interest – but one which, supposedly, contributes to the overall project being managed by the peer review cartel members.

Of course the funders and grant awarders have the major role in what science gets done, and these are all parts of an interconnected bureaucratic web of senior professional researchers. The allocation of funding, hence the direction of research and the subjects deemed acceptable, has long since been captured by the peer review cartel.

*

Even more importantly than choosing the subject matter of research, the peer review cartel has captured *the ability to define success* in solving scientific problems.

To solve a problem, the cartel of dominant scientists in a field simply *declares that the problem has been solved!*

Since peer review is now regarded as the gold standard of science, when the peer review cartel announces that a problem has been solved, then that problem has *by definition* been solved.

Since truth is no longer transcendental but internal to research then nothing more needs be said: indeed there is nothing more to say. Power is truth (in modern research).

And anyone who disagrees is not competent to have an opinion, also by definition.

*

To what does the modern 'scientist' aspire? Obviously not to discover the truth about reality. Instead, he aspires to become a member of the peer review cartel – one of the group who *allocate* 'success' in science.

In other words, the modern 'scientist' aspires to become a bureaucrat, a manager, a 'politician'. In yet other words, the modern 'scientist' aspires to power – (im)pure and simple.

However, being a modern high level bureaucrat, manager or politician is incompatible with truthfulness, and dishonesty is incompatible with science; hence being a successful modern 'scientist' is incompatible with the practice of real science.

Understanding reality

A real scientist needs to want to understand *reality* - this necessarily entails first *believing in* reality (believing that reality is real), and secondly believing that one ought to discover and describe reality (which is the specific vocation of a scientist).

*

The belief in reality is a necessary metaphysical belief, which cannot be denied without contradiction - nonetheless, in modern ruling elite culture it is frequently denied (this is called nihilism); which is why modern elite culture is unprecedented in being irrational, self-contradictory and self-destroying.

But obviously, a real scientist cannot be a nihilist - whatever cynical or trendy things he might say or do in public, in his heart he must have a transcendental belief in the reality of reality and must want to know something of it.

Thus *a real scientist cannot be a member of the modern ruling elite* – therefore, a real scientist in the modern world *must be powerless...*

*

Science also involves the metaphysical belief ('metaphysical' meaning a necessary assumption which *frames* the practice of science, and is not itself part of science) – a belief in the *understandability of nature* including the human desire and capacity to understand.

(That is, understandability at some level of approximation, *sufficient* understanding - but not necessary detailed or comprehensive understanding.).

Without this belief in the understandability of nature, science becomes an absurd and impossible attempt to find the one truth among an infinite number of possible errors.

Nonetheless, in modern elite culture, a belief in the understandability of nature and human capacity is routinely denied - another aspect of nihilism. Among many other consequences, this denial destroys the science which makes possible modern elite culture.

*

Explaining reality is a second step which *may* follow understanding, but effective explaining needs to be preceded by the desire to explain reality *accurately*, which itself entails honesty; again because there are an infinite number of *possible* explanations varying in accuracy between as close-as-possible to understood reality; to as far from accurate as you can get-away-with.

*

Modern science is undercut by many things - one is the difficulty for modern scientists of working according to the proper motivations and beliefs of a real scientist.

Transcendental beliefs such as the reality of reality and the desirability of truth are difficult to hold in isolation and in a hostile environment that imposes multiple pressures to abandon proper motivations to expedience.

It is difficult, in other words, for a modern scientist to work according to the principles of real science; when to do so requires a lesser or greater sacrifice of career and status. And when any level of sacrifice of principles will negate the possibility of real science.

Yet the demands of real science are absolute. There can be no compromise with truth.

And the punishment for failure to be truthful is simple – failure of knowledge. No progress in science – but instead loss and destruction of knowledge.

Real science declined because scientific genius declined

That science progressed overall, rapidly and by a great deal between, say, 1700 and 1950 can be assumed.

But what drove this progress?

Scientific progress is talked about in three main ways, depending on the numbers/ proportion of the population involved in generating this progress. We could conceptualize science as the product of tiny minority of creative *geniuses*, an *elite class* of professionals, or a *mass population* of competence.

*

1. Genius – science as the product of 10s to 100s of people per generation (for England at its height – much less for most other places) – a fraction of one percent of the population.

    This idea states that science is the product of, depends on, a relatively small number of geniuses - without whom there would be no significant progress.

    (The ingredients of 'genius' have been substantially elucidated by H.J Eysenck in his 1995 book of that name. These include high general intelligence (IQ), and also creativity – that is a tendency to make broad associations and connections between disparate phenomena. Eysenck notes that creativity is often an aspects of a 'masculine' personality type that is *moderately-*high in what he termed 'psychoticism'; which includes elements such as selfishness, impulsivity and a tendency to mental illness and antisocial behaviour. In other words, many geniuses are 'difficult' people...).

Therefore an age of scientific progress can be boiled down to the activity of tens or hundreds of geniuses; and the history of science is a list of great men.

Since little/ nothing is known about how to *generate* scientific genius, the task is mainly one of recognition selection of individuals; aiming to ensure that those who seem, potentially, to posses creative genius are given the chance to implement it – rather like the 'methods' for discovering and developing top athletes and sportsmen, chess grandmasters, or great singers and classical musicians.

*

2. Elite - 1000s to 10,000s of people per generation – a few percent of the population.

Science is the product of an elite of highly educated and trained people, usually found in a relatively small number of elite and research-orientated institutions, linked in an intensely intercommunicating network.

This elite are presumed to generate, by their cooperation, significant scientific progress.

Without this elite, and these elite institutions, there would be no significant progress.

According to this view, the history of science is a history of institutions. So the promotion of science is a matter of the creation and sustenance of elite degrees, elite universities, elite research units etc.

A matter, therefore, of selection of institutions.

*

3. Mass - 100,000s to millions of people per generation – a large percent of the population, ideally most of the population.

    By this view, science is the product of a 'critical mass' of scientifically-orientated and educated people spread across a nation or culture; and whose attitudes and various skills add or synergize to generate scientific progress. If society as a whole is not sufficiently 'scientific' in this sense, then there will not be significant progress.

    The history of science is seen as a history of gradual transformation of populations - mainly by educational reform. So the promotion of science is a matter of *science teaching* (e.g. in STEM – science, technology, engineering and mathematics) – to as high a level and for as many of the population as possible.

    A (common) twist on this is the idea that all humans have vast untapped potential - and that this potential might *somehow* be activated - e.g. by the right kind of education; leading to an elite of geniuses, or a mass-elite, or something...

*

Perhaps the mainstream idea nowadays is a mushy kind of belief/ aspiration that science is essentially elite but that the elite can be expanded indefinitely by education and increased professionalization.

Another common modern variant is that scientific progress *began* as based on individual creative genius, then became elite-driven, and nowadays is a mass ('democratic') movement.

However, this is merely an historical description of what has actually happened (more or less) to *professional research* - underpinned by the unchallenged (but false) assumption that scientific progress has indeed been maintained throughout this transition.

But there is no reason to accept that assumption of continued progress (given the vastly increased level and pervasiveness of hype and dishonesty in science).

Certainly there do seem to be historical examples of scientific progress without need for a prior scientific mass of the population, or even a pre-existing elite gathered in elite institutions. It looks very much as if science is mostly a product of individual genius; and a sufficient concentration and succession of creative geniuses are the key necessity - without which scientific progress will not happen.

*

Of course, nowadays there are (approximately) zero geniuses in science, so admitting that genius is necessary to significant scientific progress entails admitting that we are not making progress.

Again: *admitting that there are no geniuses means admitting there is no progress...*

which admission would devastate all scientific *careers*, since these careers depend upon the conviction and expectation of continued progress.

Therefore, the necessity for genius in science is an hypotheses that cannot be entertained.

*

Nonetheless, my reading of the history of science is that a sufficient supply of genius really is necessary to significant scientific progress (although history has not always recorded the identities of the presumed geniuses).

At any rate, science has often made significant progress without elites in the modern sense, and elites often fail to make progress; and the idea that scientific progress arises from mass education of the masses is very obviously sheer moonshine, without a shred of evidence in support...

Furthermore, if geniuses are necessary for real scientific *progress*, and if real scientific progress is necessary for *modernity* (i.e. a society based-on growth - such that growth in productivity will out-run population growth)...

And if (as it seems) there are (for whatever reason) no more geniuses...

Then scientific progress *has already stopped and will not re-start* (unless there can again be not just a few but a sufficiency of real geniuses in science) – and modern society will in due course collapse due to the usually-operative 'Malthusian' mechanism that the weight of population will grow to be in excess of economic (especially food) production.

Human capability peaked decades ago, and has since declined

What is the 'evidence' for decline in science?

Clearly, such evidence must be of the 'common sense' variety, since scientific evaluations are precisely what is under question – we know they are poisoned by dishonesty, hype and spin.

Here is one item: I suspect that *overall human capability* (leaving aside specific domains) reached its peak or plateau around 1965-75 – at the time of the Apollo moon landings – and has been declining ever since.

*

This may sound bizarre or just plain false, but the argument is simple. That business of landing men on the moon and bringing them back alive was *the supreme achievement of human capability*, the most difficult problem ever solved by humans.

40 years ago we could do it – and repeatedly – but since then we have *not* been to the moon, and I suggest the real reason we have not been to the moon since 1972 is that we cannot any longer do it. Humans have lost the capability.

*

Of course, the standard line is that humans stopped going to the moon only because we no longer *wanted* to go to the moon (done that, got the T-shirt...), or could not afford to, or something…

But I am suggesting that all this is so much hot air, merely excuses for not doing something which we *cannot* do.

*

It is as if an eighty year old ex-professional-cyclist was to claim that the reason he had stopped competing in the *Tour de France* was that he had now had found better ways to spend his time and money.

This may be true; but does not refute the fact that an 80 year old could not successfully compete in international cycling races even if he wanted to.

And this fact would not be altered if the 80 year old had undergone extensive plastic surgery and offered in evidence carefully 'airbrushed' photographs that made him look as if he was just 45.

And this fact would not be altered if he was able to do *other things instead* (such as building better computers or making better televisions).

And the fact would not be altered even if he presented the testimony of a panel of prestigious doctors and physiologists who swore on oath that he *could* win the *Tour de France* if he *really wanted to*.

He may look like he can do it, he may be able to do other things, he may swear that he could do it if he wanted to – but the telling fact is that *he does not do it*.

*

Human capability partly depends on technology. A big task requires a variety of appropriate and interlocking technologies – the absence of any one vital technology would prevent attainment.

I presume that much technology has continued to improve since 1975 – so technological decline is not likely to be the reason for failure of capability.

But, however well planned, human capability in complex tasks also depends on 'on-the-job' problem-solving – the ability to combine expertise and creativity to deal with unforeseen situations.

And human capability also depends on attitude: with the primary imperative of *getting-the-job-done*.

*

It is on-the-job problem-solving and getting-the-job-done attitudes which have declined so sharply over recent decades – declined to the point of rendering Western societies *helpless in the face of difficulties which could easily have been solved several decades ago*.

It might be asserted that these are trivial psychological factors, which could be changed if and when necessary. But it seems that these psychological factors *cannot* be discarded *even when it is necessary* – it is, after all, so much easier to deny the reality of the difficulties, simply to look the other way, do something else...

*

On the job problem-solving means having the best people doing the most important jobs.

For example, if it had not been Neil Armstrong at the controls of the first Apollo 11 lunar lander, but had instead been somebody of lesser ability, decisiveness, courage and creativity – the mission would either have failed or aborted.

If both the astronauts and NASA ground staff had been anything less than superb, then the Apollo 13 mission would have led to loss of life.

But since the 1970s there has been a decline in the quality of people in the key jobs in NASA, and elsewhere – because organizations no longer seek to find and use the best people as their ideal. They are *not even trying* to find the best people.

What do they do *instead* of trying to find the best people? All sorts of things – for example they try to be 'diverse' in various ways (age, sex, race, nationality etc).

And also the people in the key jobs, even when they are the best people, are no longer able to decide and command; due to the expansion of rules, committees and the erosion of individual responsibility and autonomy.

*

By 1986, and the Challenger space shuttle disaster, it was clear that humans had declined in capability – since the disaster was fundamentally caused by managers and committees being in control of NASA rather than individual experts.

It was around the 1970s that the human spirit began to be overwhelmed by bureaucracy (although the trend had been growing for many decades).

Since the mid-1970s the rate of progress has declined in physics, biology and the medical sciences – and some of these have gone into reverse, so that the practice of science in some areas has overall gone backwards, valid knowledge has been lost and replaced with phony fashionable triviality and dishonest hype.

Some of the biggest areas of science – medical research, molecular biology, neuroscience, epidemiology, climate research – are almost wholly trivial or bogus. They have *failed to deliver* on a truly catastrophic scale.

Never have so many resources have been expended with so little to show for it: Stonehenge and the Pyramids may not *do* much, but at least they are still *there...*

*

This broad general failure in core objectives is not compensated by a few islands of progress, e.g. in computerization and the invention of the internet.

Capability must cover all the bases, psycho-social as well as technical, and depends not on a single advanced area but all-round advancement in all necessary areas.

Human capability then and now

The fact is that human no longer do - *can* no longer do - many things we used to be able to do: land on the moon, swiftly win wars against weak opposition and then control the defeated nation, secure national borders, discover 'breakthrough' medical treatments, prevent crime, design and build to a tight deadline, educate people so they are ready to work before the age of 22, suppress piracy on the high seas...

50 years ago Western societies would aim to have the smartest, best trained, most experienced and most creative people they could find (given human imperfections) in position to take responsibility, make decisions and act upon them in pursuit of a positive goal.

That is what they were *trying* to do.

Now, we are *not even trying*.

And since we are not even trying to do the job, naturally the job will not be done.

*

Now we have dull and docile committee members chosen partly with an eye to affirmative action and partly to generate positive media coverage, whose major priority is not to do the job but to avoid personal responsibility and prevent side-effects and to build careers; pestered at every turn by an irresponsible and aggressive media and grandstanding politicians out to score popularity points; all of whom are hemmed-about by vast and proliferating regulations, such that – whatever they *do* do, or do *not* do, whether they succeed or fail – they will be in breach of some rule or another and vulnerable to open-ended sanctions.

*

So we should be honest about the fact that human do not anymore fly to the moon because humans cannot anymore fly to the moon.

Also noteworthy is that the deepest manned ocean descent of about 10.9 kilometres into the Mariana Trench, was as long ago as 1960; and humans have never again been as deep during the past half century.

Humans have failed to prevent or suppress the re-emergence of high seas piracy on a large scale because we nowadays cannot do it - although humans solved the problem 150 years ago.

And we cannot solve *new* problems either, since these require a combination of attitudes and freedoms that we can no longer imagine, or which we fear more than the problems themselves. In the past the average experts were both smarter and more creative than we are now, and these experts would then have been in a position to do the needful.

Measuring human capability: Moonshot versus 'Texas Sharpshooter'

But is the Moonshot really a valid measure of human capability?

Yes. The reason that the Moonshot is a valid measure of human capability is that the problem was difficult and was not chosen but imposed.

*

The objective of landing men on the moon (and bringing them safely back) was not chosen by scientists and engineers as being something already within their capability – but was a problem imposed on them by politicians.

The *desirability* of the Moonshot is irrelevant to this point. I used to be strongly in favour of space exploration, now I have probably turned against it – but my own views are not relevant to the use of the Moonshot as the ultimate evidence of human capability.

Other examples of imposed problems include the Manhattan Project for devising an atomic bomb – although in this instance the project was embarked upon precisely because senior scientists judged that the problem could possibly, maybe probably, be solved; and therefore that the US ought to solve it first before Germany did so.

But, either way, the problem of building an atomic bomb was also successfully solved.

Again, the desirability of atomic bombs is not the point here – the point is that it was a measure of human capability in solving difficult imposed problems.

*

Since the Moonshot, there have been several difficult problems imposed by politicians on scientists that *have not been solved*: such as finding a 'cure for cancer' (or the common cold) and 'understanding the brain'.

These two problems had vastly more monetary and manpower resources (although vastly less talent and creativity) thrown at them than was the case for either the Moonshot or Manhattan Project.

But modern technological advances are *not* imposed problems; they are instead examples of the Texas Sharpshooter fallacy.

*

The joke of the Texas Sharpshooter is that he fires his gun many times into a barn door, then draws a target over the bullet holes, with the bulls-eye over the closest cluster of bullet holes.

In other words the Texas Sharpshooter makes it look as if he had been aiming at the bulls-eye and had hit it, when in fact he drew the bulls-eye *only after he took the shots*.

Modern science and engineering is like that. People do research and development, and then proclaim triumphantly that whatever they have done is a breakthrough. They have achieved whatever-happens-to-come-out-of-R&D; and then they spin, hype and market whatever-happens-to-come-out-of-R&D as if it were a major breakthrough.

74

In other words, modern R&D triumphantly solves a *retrospectively designated problem*, the problem being generated to validate whatever-happens-to-come-out-of-R&D.

*

The Human Genome Project was an example of Texas Sharpshooting masquerading as human capability.

Sequencing the human genome was not a matter of solving an imposed problem, nor any other kind of real world problem, but was merely doing *a bit faster* what was already happening.

*

Personally, I am no fan of Big Science, indeed I regard the success of the Manhattan Project as the beginning of the end for real science.

But those who are keen that humanity solve big problems and who boast about our ability to do so; need to acknowledge that humanity has apparently become much worse, not better, at solving big problems over the past 40 years – so long as we judge success only in terms of solving imposed problems which we do not already know how to solve, and so long as we ignore the trickery of the many Texas Sharpshooters among modern scientists and engineers.

The Texas Sharpshooter society
of secular modernity

As I said, the Texas Sharpshooter fallacy is a joke which suggests that the TS fires his gun many times into a barn door, then afterwards draws a target over the bullet holes.

But the sharpshooter fallacy is nowadays unavoidable and everywhere, it characterizes secular modern society throughout, because secular modern society *has no aim* but instead idealizes process and *retrofits aim to outcome*.

Indeed, the Texas Sharpshooter strategy is the master theory of our phase of late modernity – the persuasion of people that whatever has happened is what they wanted and what was intended.

*

Secular moderns - in public discourse - 'believe in' things like freedom, or democracy, or equality, or progress - but these are processes, not aims.

Aims are not prescribed in advance and progress checked-against them – instead, *aims are retrospectively ascribed to whatever emerges from process.*

In this respect professional science is merely a typical aspect of modern life – real science has been assimilated into mainstream contemporary life.

*

It happens all the time: abolition of slavery emerged from the American Civil War therefore people retrospectively ascribe liberation as its purpose. Destruction of the death camps emerged from the second world war, so the liberation of the Jews is ascribed as its purpose.

Libertarians 'believe in' freedom not as a means to some end, but as a process which by definition leads to the best ends; so that they 'believe in' whatever comes out of the process.

*

The modern attitude is that the best thing is for science to be well funded and to do what science does, and whatever comes out of the process of science is retrospectively defined as 'truth'.

In practice, science is defined as *whatever scientists do*, and what scientists do is defined as generating truth.

Texas Sharpshooter Fallacy...

*

Or law. Law is a process, and justice is defined as that which results from the process of law. Modern laws may feel revoltingly unjust; but lacking a transcendental concept of justice, nothing more can be said. Justice is what justice does.

TSF...

*

Or education. What is education? The answer is 'what happens at school and college'. And whatever happens at school and college is what counts as education. Since what happens at school and college changes, then the meaning of education changes. But since education is *not aiming at anything in particular*, it is merely 'what happens at schools and colleges', then these changes cannot be evaluated. Whatever happens is retrospectively defined as what needed to happen.

TSF...

*

Or economics. Economic 'growth' is pursued as the good, and whatever comes out of economics is defined as prosperity. What people 'want' is known only by what they get - their wants are retrospectively ascribed. If what is being measured and counted grows, then this is defined as growing prosperity. So the economy fifty years ago wanted more A, B and C but the modern economy instead provides X, Y and Z – however, economists retrospectively re-draw the target around X, Y and Z and proclaim the triumph of economics. The economy did not provide ABC, but this is taken to prove that ABC was not *really* wanted; instead the economy provided XYZ which is taken to *reveal* people's *true* preferences.

TSF...

*

This is, of course, paradoxical; but it is not just paradoxical - it is nonsense.

The primacy of process is simple nonsense – it is sleight-of-hand, it is bait-and-switch. It is trying to do without aims because all aims point to the necessity for underpinning justifications for those aims. Since modern society regards clear and explicit aims as merely arbitrary and subjective statements, and because aims (except when platitudinous) are divisive; it cannot agree on aims and regards it as dangerous to try.

Secular modernity is *fundamentally* (not accidentally, not reformably) based on the Texas Sharpshooter fallacy, and the fallacy is simple and obvious

However, since the fallacy is intrinsic and pervasive, it must be concealed; and it is concealed.

Since collapse happened to Classics, it could happen to science

Since professional science is not longer providing the breakthroughs in efficiency that are necessary to sustain modernity, then modernity will collapse; we will, in other words, return to the Malthusian Trap in which increasing population will cause reducing standard of living (or violence or disease) until such a point that the population has come into line with resources.

But, before that point, it is probable (not definite) that professional science will itself collapse – simply because it is on the one hand a waste of resources (costs) and on the other hand these resources are *needed for other purposes* (opportunity costs).

*

I find that people simply cannot take seriously that Science would collapse down to a small fraction of its current (vast, bloated) size. Despite that real science is so recent a feature of human history, and is so fragile (so vulnerable to corruption) – people assume that it must be eternal because it so *useful*.

But why people imagine that something will survive merely because it is useful, in the face of so many counter examples in their own experience of useful things disappearing, is hard to fathom...

And there is a recent precedent for the collapse of the dominant intellectual culture: Classics.

*

The study of Greek and Roman culture - language, history, literature, philosophy - was the dominant secular intellectual activity in the West for many hundreds of years and the teaching of Latin was at the core of the educational curriculum for a couple of millennia.

Latin was the mark of A Gentleman, especially A Scholar – Classics was the highest status form of knowledge, the main (sometimes he only) subject taught at the best schools and universities.

*

In England, when it was the top country and culture, Classics pretty much monopolized the curriculum in the Public Schools, Grammar Schools and Oxford University (Cambridge focused on mathematics - but had plenty of Classicists too). New subjects like Science, modern languages and modern history had to fight for space in the curriculum.

Right up into the mid 20th century, the most prestigious general degree in England was the Oxford four year Classics degree – it was the premier 'qualification' for elite ruling class professions.

The 'two cultures' debate of the late 1950s and early 1960s marked the tipping-point when Science began to dominate Classics in general cultural discourse. Classics more-or-less retained prestige for another generation, after which it very suddenly collapsed, in the 1980s.

*

The classics have now dwindled to the status of a hobby, taught in few schools and very seldom given much prominence.

Most UK universities have all-but abandoned the subject except at a 'taster' level - only a handful of courses at a few places can find undergraduates with any background or competence in Latin (even fewer in Greek); so most modern 'Classics' degrees are built on no foundations in three years; teaching from a basis of zero knowledge.

Advocates of Classics find it ever harder to justify their subject as worthy of study - certainly there is no automatic deference towards it, no assumption of its superiority.

*

So, in the space of about 250 years, from the time of Samuel Johnson - when he was apologetic about writing in English rather than Latin and focusing his dictionary on the English vernacular - until now, Classics have dwindled from unchallenged dominance to insignificance in general Western culture.

*

Classics was quietly dwindling in cultural importance for a few hundred years (at least since Shakespeare outstripped all rivals using the vernacular), and this was becoming ever more apparent from the mid 19th century; but at least as recently as the time of the great English Classics professor (and poet) Houseman (1859-1936) it looked as if the subject was on the verge of a breakthrough (using 'modern' scholarship).

And of course classical scholarship has continued throughout all this decline, pouring-out research books and scholarly articles for a dwindling audience of other scholars.

But despite all this, Classics has undeniably collapsed.

*

My point is that if it seems unimaginable to many people that Science really could collapse from dominance into insignificance in just a few decades, then these people should think about what happened to Classics. The signs are there for those who look behind the hype.

Of course a scientist feels that the real importance of Classics was trivial compared with Science – that the modern world *depends on Science*.

Quite true; but then the ancient world depended on Classics, and the collapse of Classics was linked with the collapse of traditional society.

The collapse of Science is linked with the collapse of modernity - both as cause and as consequence.

Chargaff on the loss of human pace and scale in science

Referring to his first twelve years at Columbia University, USA, Erwin Chargaff (1905-2002) said:

"The more than sixty regular papers published during that period dealt with a very wide field of biochemistry, as it was then understood; and a few of them may even have contributed a little to the advance of science, which, at that time, was still slow, i.e., it had human proportions... Nevertheless, when I look back on what I did during those twelve years, there come to mind the words ascribed to St. Thomas Aquinas: *Omnia quae scripsi paleae mihi videntur.* All he had written seemed to him as chaff.

"When I was young, I was required – and it was easy – to go back to the origins of our science. The bibliographies of chemical and biological papers often included reference to work done forty or fifty years earlier. One felt oneself part of a gently growing tradition, growing at a rate that the human mind could encompass, vanishing at a rate it could apprehend.

"Now, however, in our miserable scientific mass society, nearly all discoveries are born dead; papers are tokens in a power game, evanescent reflections on the screen of a spectator sport, new items that do not outlive the day on which they appeared. Our sciences have become forcing houses for a market that in reality does not exist, creating, with the concomitant complete break in tradition, a truly Babylonian confusion of mind and language.

"Nowadays, scientific tradition hardly reaches back for more than three or four years. The proscenium looks the same as before, but the scenery keeps on changing as in a fever dream; no sooner is one backdrop in place than it is replaced by an entirely different one. The only thing that experience can now teach is that it has become worthless.

"One could ask whether a fund of knowledge, such as a scientific discipline, can exist without a living tradition. In any event, in many areas of science which I am able to survey, this tradition has disappeared. It is, hence, no exaggeration and no coquettish humility if I conclude that the work we did thirty or forty years ago – with all the engagement that honest effort could provide – is dead and gone."

Erwin Chargaff – *Heraclitean Fire*, 1978.

*

From this I note: "the advance of science … was still slow, i.e., it had human proportions. … One felt oneself part of a gently growing tradition, growing at a rate that the human mind could encompass, vanishing at a rate it could apprehend."

That is the pace of real science.

"…in our miserable scientific mass society, nearly all discoveries are born dead; papers are tokens in a power game, evanescent reflections on the screen of a spectator sport, new items that do not outlive the day on which they appeared…"

In contrast, the "miserable scientific mass society" of modern research does not operate at the pace of real science, but at the pace of *management*.– Six monthly appraisals, yearly job plans, three yearly grants and so on. All evaluations being conducted and determined by committee and bureaucracy, by votes and algorithms, according to check-box lists of objectives and outcomes - rather than by individual judgment.

"Our sciences have become forcing houses for a market that in reality does not exist…"

Nobody really *wants* what modern science provides, there is no real *need* for it; which is why modern science is dishonest – from top to bottom: modern science must engage in public relations, hype, spin – lies – in order to persuade 'the market' that it really wants whatever stuff the 'forcing houses' of modern science are relentlessly churning-out.

"…honest effort…"

A two-word definition of real science.

Delbruck on the moral qualities
of science

Max Delbruck - 1906-1981. Nobel Prize 1969

Question: Does scientific research by itself foster high moral qualities in men?

Delbruck's answer: "Scientific research by itself fosters one high moral quality: that you should be reasonably honest. This quality is in fact displayed to a remarkable extent. Although many of the things that you read in scientific journals are wrong, one does assume automatically that the author at least believed he was right."

(Quoted p282 in *Thinking about Science: Max Delbruck and the origins of molecular biology*. EP Fischer & C Lipson. 1988)

*

Delubruck was talking in 1971, forty years ago (a *mere* 40 years ago!) and he was one of the most well-connected of twentieth century scientists, a kind of godfather to molecular biology, and a man of great personal integrity.

So Delbruck was in a position to know what he was talking about.

And, in 1971, he was able to state that scientific research by itself fosters the high moral quality that you should be reasonably honest. And that this quality is *in fact* displayed to a remarkable extent.

And that when reading journals scientists could and did *assume that the authors were telling the truth as they saw it.*

Only 40 years ago Delbruck could state that scientists were in fact, in reality, in practice - honest...

Nobody of Delbruck's integrity in Delbruck's position could say the same today.

Micro-specialization and the infinite perpetuation of error

Science, real science, is itself a specialization of philosophy. After which science itself specialized – at first into physical and natural sciences, and then into ever-finer divisions.

Scientific specialization is generally supposed to benefit the precision and validity of knowledge *within* specializations, but at the cost of these specializations becoming narrower, and loss of integration *between* specializations.

In other words, as specialization proceeds, people supposedly know more and more about less and less - the *benefit* being presumed to be more knowledge within each domain; the *cost* that no single person has a general understanding.

*

However, I think that there is no benefit, but instead harm, from specialization beyond a certain point – an imprecise but long-since-passed point.

Nowadays, people do not really *know* more, even within their specialization – often they know nothing valid at all; almost everything they think they know is wrong, because undercut by fundamental errors *intrinsic and yet invisible to* that specialty.

The clear cut benefits of specialization apply *only to the early stages* such as the career differentiation in the early 20th century - the era when there was a threefold division of university science degrees into Physics, Chemistry and Biology.

It is much less obvious that real science benefited from subdivision of each of these into two or three (e.g. Physics into Theoretical and Applied, Chemistry into Organic and Inorganic; Biology into Zoology and Botany).

But since the 1960s scientific specialization has now gone far, far beyond this point, and the process is now *almost wholly disadvantageous*.

We are now in an era of *micro-specialization*, with dozens of subdivisions within sciences. Biology, for example, fragmented into biochemistry, molecular biology, genetics, neuroscience, anatomy, physiology, pharmacology, cell biology, marine biology, ecology...

*

Part of this is simply the low average and peak level of ability, motivation and honesty in most branches of modern science. The number of scientists has increased by more than an order of magnitude – clearly this has an effect on quality.

Scientific training and conditions have become prolonged and dull and collectivist – deterring creative and self-motivated people. And these changes have happened in an era when the smartest kids tended not to gravitate to science, as they did in the early 20th century, but instead to professions such as medicine and law, and into the financial sector.

In round numbers, it seems likely that more than ninety percent of modern 'scientists' are worse than the worst scientists of 60 years ago.

However there is a more basic and insoluble problem about micro-specialization. This is that micro-specialization is about micro-validation – which can neither detect nor correct gross errors in its basic suppositions.

*

In the world of micro-specialization that is a modern scientific career, each specialist's attention is focused on technical minutiae and the application of conventional proxy measures and operational definitions. Most day-to-day research-related discussion (when it is not about fund-raising) is *troubleshooting* – getting techniques and machines to work, managing personnel and coordinating projects...

Specific micro-specialist fields are built-around specific methodologies - for no better ultimate reason than 'everybody else' does the same, and (lacking any real validity to their activities) there must be some kind of arbitrary 'standard' against which people are judged *for career purposes* (judging people by real scientific criteria of discovering truths is of course *not done*).

('Everybody else' here means the cartel of dominant Big Science researchers who control peer review - appointments, promotions, grants, publications etc. - in that micro-speciality.)

Thus, micro-specialists are ultimately technicians and/or bureaucrats; thus they cannot even *understand* fatal objections and comprehensive refutations of their standard paradigms when these originate from adjacent areas of science. So long as their own specific technique has been conducted according to prevailing micro-specialist professional practice, they equate the outcome with 'truth' and assume its validity and intrinsic value.

In a nutshell, micro-specialization allows a situation to develop where the whole of a vast area of science is bogus knowledge; and for this reality of total bogosity to be intrinsically and permanently invisible and incomprehensible to the participants in that science.

*

If we then combine this situation with the prevalent professional research notion that *only micro-specialists are competent to evaluate the domain of their micro-speciality* – and add-in the continual fragmentation of research into ever-smaller micro-specialties - then we have a recipe for *permanent and intractable error.*

*

Vast and exponentially-growing scientific enterprises have consumed vast resources without yielding any substantive progress at the level of in-your-face common sense evaluations; and the phenomenon continues for time-spans of whole generations, and there is no end in sight (short of the collapse of science-as-a-whole).

According to the analysts of classical science, science was supposed to be uniquely self-correcting - in practice, now, thanks in part to micro-specialization, it is not self-correcting *at all* – except at the trivial and misleadingly reassuring level of micro-defined technical glitches and slip-ups.

Either what we call science nowadays is not 'real science' or else real science has mutated into something which is a mechanism for the perpetuation of error.

The idea of science as a *truth-machine*

As I survey the wondrous corruption of scientists, it seems that *most* of them fall pretty soon into the obvious dishonesty of hype and spin, selection and exaggeration.

But not all do so; and among those decent scientists who strive to be honest while pursuing a successful career I perceive an alternative pattern which is only indirectly, and as it were accidentally dishonest: a pseudo-solution which unintentionally makes matters worse, by camouflaging flagrant dishonesty and rejecting real science.

It is a strategy which is often pursued with high ideals, a clear conscience and in a spirit of modesty, although it is in operation anti-scientific in spirit and effect.

This strategy is to replace honesty with precision, to replace truth-seeking with a quest for technical accuracy.

*

Perhaps the root of this error is the notion that there is such a thing as 'scientific method' (detachable from the individuals who practice science); and that *if* this scientific method is strictly adhered-to, *then* the result will be valid science.

In parody, this is the terribly mistaken view that *science is a truth-machine*: the idea that *if* you do science properly *then* you will manufacture 'truth' reliably and cumulatively.

The idea that *if* you perform observations and experiments according to the approved principles, *then* this will lead to 'facts'. And *if* you feed these 'facts' into the correct analytical and statistical procedures ('scientific methodology') *then* what comes out of the machine will be objective truth.

The idea that although the exact output may not precisely be known in advance, the *process* by which valid outputs are generated is understood to be *controllable*, and *therefore* it can confidently be predicted that the result of this process will be valuable knowledge.

In sum, this is the mainstream modern view that research *input* reliably leads to real scientific *output* (albeit with varying degrees of efficiency).

(This reasoning justifies the usual practice of measuring science by measuring *inputs* – that is, measuring science by measuring how many resources - grants, personnel, capital - are expended on supposedly-scientific *goals*. The inputs are simply *assumed* to result in valid and relevant outputs of knowledge so long as approved procedures are strictly followed.)

*

This is, indeed the basic underlying 'model' for modern science, especially Big Science – and it leads to the mainstream assumption that the constraint on science is resources. The model assumes that – if you have research managers who are deploying resources (manpower, machines etc) doing the right things - then resources will be transformed into knowledge.

There may be disagreement about the *efficiency* of this process, but the assumption is very widely held that spending a lot of money on a problem *will* accumulate knowledge towards its solution – so long as the researchers are competent and rigorous (and that

94

competence and rigour are themselves defined as products of resources – i.e. educational and training resources).

Indeed, *rigour* is a key word here – because rigour is defined in term of exact adherence to predetermined method, technique, procedure – and this implies that science ideally ought to be made wholly explicit, planned down to its finest detail, and done in accordance with these plans.

And this is, indeed, the way that research funding is managed – 'scientists' are compelled to submit detailed plans, which are approved or disapproved. And research performance is monitored against these plans.

Science is seen as a process of *implementation*, the process is seen as something explicit and managed, and the role of the individual researcher is – in a nutshell – *obedience*.

*

It is obvious that this typically modern way of doing professional research – based on the concept of research as a 'truth-machine' - bears zero relationship to how real science was done in the past, during the golden age of science - when science was small scale, individualistic, cheap, efficient and led to many breakthroughs; and it is also obvious that this resource- and organization-orientated way of doing research is derived not from science but is instead the characteristic *modus operandi* of bureaucracies.

And it is worth asking what evidence there is or was that scientific research would be done better, valid knowledge better generated, in this bureaucratic fashion than in the effective mode of the past?

And the answer is equally obvious that there is *no such evidence*; but, on the contrary, conclusive evidence that scientific research is done *much worse or not at all* with this bureaucratic mode – less efficiently and less effectively, indeed mostly done with zero or negative real world outcomes – than when research was done as real science.

*

The deep problem with a technical focus on rigour is that *method is a means not an end.*

A scientific problem does not dictate a specific method; indeed the solution to a problem often comes from a new, non-obvious and unanticipated method; and the solution to a problem is often best known exactly by the convergence of several methods.

Furthermore, methods are substantially constrained by manpower and technology (especially by the development of machines, including computers), and a focus on method becomes a race to assemble the largest teams and be the first to deploy the new and 'improved' technology. Linked is the assumption that old technology and methods are intrinsically *unable* to answer the questions. Whereas old technologies and methods may well be able to answer the questions *if* creative scientific genius is added to the mix – or even just sincere truth-seeking.

*

Yet, as often as not, modern scientific 'fields' (research groups, appointments, journals, conferences) are defined by their technologies. Presumably this is helpful in terms of proximate goals - trouble-shooting methods; but destroys the possibility of real science.

Real science does not happen unless scientists are truth seekers and truth speakers, and truth seeking is an end not a means – truth is not confined by methods; and truth is a whole, not a part – an excess focus on one aspect is equivalent to the gross exaggeration of one virtue at the expense of virtuous-ness.

So factual technical rigour, being measurable, on the one hand obscures all other forms of dishonesty– so that people who are *not even trying* to discover or tell the truth but instead pursuing full-on careerism can nonetheless feel themselves to be self-denying paragons of virtue due to their slavish and uncritical submission to fashionable but arbitrary technical demands; while on the other hand it also rejects real science as being insufficiently rigorous in terms of its having lower precision in the approved technical domain.

And, because science has been fragmented into micro-specialties, these errors are *ineradicable* – there is no progress through self-correction, merely the fashion-driven progression of new techniques.

Zombie science

Since modern researchers have abandoned the core ethic of truth seeking, most 'scientists' are quite willing to pursue wrong ideas so long as they are rewarded for doing so with sufficient career incentives.

The primary criterion of the 'validity' of a modern research field is therefore, in practice, the probability that working in it will be likely to benefit your career.

*

Nowadays, when a new idea is launched, it is unlikely to win converts unless early-adopters are rewarded in an upfront and obvious fashion – typically with incentives such as research funding, the opportunity to publish in prestigious journals, and the promise of increased status exemplified by interest, admiration and respect from other researchers.

This is the currency of science – the tokens used to exchange for status, jobs, salary, promotions, prizes...

Therefore new research fields and theories may *with extraordinary rapidity* become popular and even dominant purely and simply because adoption is reinforced by career incentives.

Scientific strengths or scientific weaknesses are nowadays strictly irrelevant.

*

In terms of the classical theory of science; worthless theories (e.g. theories that are incoherent or fail to predict observations) should be demolished by sceptical (or jealous) competitor scientists, who will denounce the weaknesses of merely-fashionable theories in person, in conferences and (especially) in print – in the scientific record, the 'literature'.

However, in practice it seems that even the most conclusive 'hatchet jobs' done on phoney theories will fail to kill, or even weaken, them - when the phoney theories are backed-up with sufficient career incentives. Scientists gravitate to where the money is; and the paraphernalia of specialist conferences (to present results at), journals (to publish in) and academic jobs (to work in) will follow the money as night follows day; so long as the funding stream is sufficiently strong, deep and sustained.

Classical scientific theory has it that a wrong hypothesis will be rejected when it fails to predict 'reality' as determined by controlled observations and experiments. But such a catastrophe can be deferred almost indefinitely by the elaboration of secondary hypotheses to explain why failing to fit the facts is not – after all – fatal to the theory; but instead merely implies the need for a more complex theory – which then requires further testing, and extra funding, and generates more paid work for the bogus believers.

Furthermore, since the new version of the bogus theory, with its many auxiliary secondary hypotheses, is so complex – this complexity makes it that much harder to test; indeed conclusive tests may become impossible, even in principle, since there are no precise predictions. All of which has the effect of putting-off indefinitely the time when a bogus theory needs to be abandoned.

(Meanwhile, a much simpler rival theory – i.e. that the first theory is phoney, and always was phoney, and this is why it so singularly fails to predict reality – is regarded as simplistic, crass, merely a sign of lack of sophistication ...)

*

After a while, lavish funding creates powerful interest groups associated with the phoney theory - including the reputations of numerous scientists who are now successful and powerful on the back of the phoney theory, and who by-now control the peer review process (including allocation of grants, publications and jobs) so there is a powerful disincentive against upsetting the apple cart. Indeed, the system of peer review sustains the phoney theory on the basis that turkeys do not vote for Christmas.

What is the function of Zombie science?

When a branch of science based on incoherent, false or phoney theories is serving a useful *but non-scientific* purpose it may be kept-going by continuous transfusions of cash from those whose non-scientific interests it serves.

Such an interest group moves yet is not truly alive, acts with intent yet cannot be killed; it is an un-dead creature without sentience, externally animated by funding: a Zombie science.

For example, if a branch of pseudo-science based on a phoney theory is nonetheless valuable for political purposes (e.g. to justify a government intervention such as a new tax or subsidy) or for marketing purposes (to provide the rationale for a sales campaign) then real science expires and a 'zombie science' evolves.

Zombie science is science that is *dead but will not lie down*. It keeps twitching and lumbering around so that (from a distance, and with your eyes half-closed) zombie science looks much like real science. But in fact the zombie has no life of its own; it is animated and moved only by the incessant pumping of funds.

*

Real science is coherent – and testable (testing being a matter of checking coherence with the result of past and future observations).

Real science finds its use, and gets its validation, from common sense evaluation and being deployed in technology.

Real science is validated (contingently) insofar as it leads to precise predictions that later come true; and leads to new ways of solving pressing problems and making useful changes in the world.

But zombie science is not coherent, therefore cannot be tested; its predications are vague or in fact retrospective summaries rather than predictions.

*

In a nutshell, zombie science is supported because it is *useful propaganda*; trading on the prestige which real science used-to have and which zombie science falsely claims for itself.

Zombie science is deployed in arenas such as political rhetoric, public administration, management, public relations, marketing and the mass media generally. It persuades, it constructs taboos, it buttresses rhetorical attempts to shape opinion.

Furthermore, most zombie sciences are supported by moral imperatives – to doubt the zombie science is therefore labelled as wicked, reckless, a tool of sinister and destructive forces.

To challenge zombie science is not merely to attack the livelihoods of zombie scientists (which, considering their consensus-based power, is itself dangerous) – but opens the attacker to being labelled a luddite, demagogue, anti-science, a *denialist!*

For all its incoherence and scientific worthlessness, zombie science therefore often comes across in the sound bite world of the mass media as being *more plausible than real science*; and it is precisely the superficial face-plausibility which in actuality is the sole and sufficient purpose of zombie science.

The expectation of growth in scientific knowledge

We have become used to growth in scientific knowledge, and have come to expect growth in scientific knowledge – scientific progress.

This expectation in scientific progress at first shaped reality and eventually displaced reality.

The link between expectation and actuality was broken and the world of assumptions took over.

Enter Zombie science.

*

The expectation that scientific knowledge will grow almost inevitably (given adequate 'inputs' of personnel and funding) is epitomized by the professionalization of scientific research (making scientific research a career, then making research jobs part of a bureaucracy) and the normal career expectation of regular and frequent and measurable *outputs* – especially research publications.

The expectation of regular and frequent research publication would only make sense if it was assumed that scientific knowledge was accumulating in a predictable fashion.

*

Because of what happened in the past, we nowadays expect an open-ended growth in the number of scientific publications over time, and a growth in the totality of citations (references to previous research).

These quantitative increases are at bottom fuelled by increases in the *numbers of professional scientists* which drives the number of journals for publishing science.

According to analysis by Michael Mabe, each researcher generates approximately one-paper-per-year (controlling for number of authors per paper); and this has not changed significantly over the decades.

Leaving aside quality – it seems that the expansion of scientific publication and the growth in journals is merely a consequence of the expansion of manpower.

*

By assuming that growth in researchers and publications will continue, we implicitly assume that there is an unbounded quantity of new and useful science waiting to be discovered and an unrestricted pool of people capable of making discoveries.

The economist Paul Romer – and others – built this into theories of the modern economy – the argument is that continued growth in science and technology fuels continual improvement in productivity (economic output per person) and therefore growth in the economy. Some of this economic growth is invested into science (education and employment of personnel and capital equipment) to drive further economic growth.

The idea is: Increased science leads to increased productivity leads to increased science.

The idea is that we are *continually getting better at scientific discovery*, because we are continually investing more in scientific discovery, *therefore* modern society is enabled to continue economic growth – pretty much *forever...*

*

Plausibility aside, how would we know whether *real* science was growing? (Useful new science, that is, as contrasted with the mere volume of publications and other communications – which is simply a matter of words, pictures and numbers.)

Who could evaluate whether change is *real* science not hype; and whether increased amounts of self-styled scientific *stuff* (publications, personnel, laboratories etc.) actually corresponded to more and better real science?

Why *should* the default assumption be that increased size of science as a professional activity corresponds to increased knowledge?

*

When scientific growth is expected, and when society acts-upon the expectation, we have an overwhelming assumption of growth in science, an assumption that science is growing – but that assumption says *nothing at all* about whether there really *is* growth.

Indeed, so strong is the assumption that science is progressing that we have a situation where a critic of science is expected to prove the negative - that science is *not* growing. This is remarkable – instead of scientists and the funders of science having to prove that *they are making progress* we have a situation where the progress is assumed (unless proven otherwise).

We have a situation where every scientific publication (once it has been published) is nowadays presumed to be valid (unless proven otherwise).

This was not always the case! – to put it mildly.

*

One reason people assume an increase of knowledge is that modern science has been 'peer reviewed' before publication – that is to say a virtual-committee of other scientists have – if not exactly approved it, then at least not vetoed its publication, based on their opinions of its quality.

However, what the opinion of a bunch of self-styled scientists has to do with validity is never explained – since real science was *supposed* to be about something more than opinion; indeed being about more than mere opinion was *precisely* what supposedly made science *science*!

Yet since we are in a situation when, apparently, published science is assumed to be valid (until proven otherwise), and in a situation when publication depends on an opinion poll (i.e. peer review) – then 'science' of this kind is no more valid than any other opinion poll.

*

The situation is that a bunch of senior researchers in each research specialty (the peer review cartel) are assuring the outside-world that *yes* – research volume is in fact the same as knowledge volume. And since research volume is a product of research input, the conclusion is obvious: *give us more money!*

*

When people *assume* real science is growing, and when they think they perceive from output volume that real science is growing, this creates vast possibilities for dishonesty, hype and spin.

Because people are expecting science to grow, and expecting there to be *regular breakthroughs*, they tend to believe it when regular breakthroughs are *claimed* (whether or not breakthroughs have actually happened).

Indeed, if the breakthroughs are not obvious – then breakthroughs will be looked for until breakthroughs are 'found'.

(The same happens with genius – if there are no modern geniuses to compare with the scores of geniuses observable 100 years ago – then this is assumed to mean that we need to look harder for the geniuses which we just *know* must be there...)

*

But how if there is really no growth in scientific knowledge – but merely growth in scientific communications?

How if there is actual *decline* in real scientific knowledge – how would we know this from outputs?

We wouldn't. We would only know it from declining *capability*. And declining capability, as argued above, we *do* seem to observe.

Doing real science is *hard*

A thought experiment: Let us suppose that doing real science is actually much *harder* than most people currently assume; much slower, more difficult and less predictable.

*

Suppose that *most* competent and hardworking real scientists actually make *no indispensable and distinctive contribution to real science* – none at all - but merely incremental improvements, minor checks or refinements in methods, the precision of measurements and the expression of theories.

And if they personally had not done it, it would have slightly-slowed but would not have prevented progress: or somebody else would have done it.

And that it is simply the nature of real science that unpredictability, riskiness, is *intrinsic*. That there are no guarantees that even able and hardworking scientists will actually achieve *anything at all*.

Indeed suppose that the likeliest outcome is that most scientists – most *real* scientists - will turn-out to have been redundant, superfluous, and unnecessary – yet this outcome can be known only in retrospect. We cannot predict who will succeed and who will fail.

And of course all this is *not* a thought experiment – it is a simple statement of fact: *real science is really hard.*

(Ask any real scientist.)

*

Since real science is really hard, then this fact is *incompatible with the professionalization of science* – with the idea of scientific research as a career.

Since real science is irregular and infrequent and unpredictable, then science could only be done in an *amateur* way; maybe as a sideline from some other profession like teaching, practicing medicine, being a priest or skilled craftsman, or as a hobby for the wealthy – or supported by a patron.

Professional science would then be regarded as an activity *intrinsically phoney*; and the phoniness would increase as professionalization of science increased and became more precisely measured, and as the profession of science expanded – until it reached a situation where the visible products and evidence of science (publications, personnel, buildings, organizations and their funding) – the *stuff* bore zero relationship to the reality of science.

Professional scientists would produce stuff (like scientific publications) regularly and frequently, but this stuff would actually have nothing to do with real science.

Nothing to do with real science.

*

Or, more exactly, the growing amount of stuff produced by the growing numbers of professional science careerists, whose use of hype would also be growing – the amount of *this* stuff would be *so much greater* than the amount of real science, that any *real* science would be obscured utterly.

*

This is precisely what we have.

The observation of growth in scientific knowledge became an expectation of growth in science and finally an assumption of growth in science.

And when it was being assumed that science was growing, the amount of real science did not really *need* to grow, because the assumption that it was growing framed the reality.

*

But if real science is as *hard* (slow, unpredictable, uneven) as was always believed (until the mid 20th century); then scientific progress cannot be taken for granted, cannot be expected or assumed.

Yet our society *depends* on scientific progress – if or when scientific progress stops, our society will soon collapse, because it will not grow – and our society depends on growth.

Despite, or maybe because, science is vital to our survival as a civilization, so great is our societal arrogance that we do not regard science as something *real*.

Instead we have made scientific progress a matter of casual assumption, not of serious observation; hence scientific progress has become the subject of unbounded wishful thinking and deceitful propaganda.

When the bubble bursts

Real science is a way of getting at certain kinds of truth, but the way that science works is absolutely dependent on honesty and integrity. Our societal arrogance is such that we believe that we can have the *advantages* of real science but at the same time *subvert* the honesty and integrity of science whenever that happens to be expedient.

We act as if real science can necessarily be formalised, mechanised and made a process of mass production. And we don't even attempt to check whether this is *true*.

Our societal arrogance is that we are in control of this dishonesty – that the amount of hype and spin we apply to science is under our control, trivial in its effect, and can be undone at will; that we can separate the signal of honest real science from the noise of mass produced 'research'- and 'back-calculate' or reverse-engineer the truth of science from the lies and exaggerations of careerist research...

*

But in fact we have *no idea* of the real situation in science, no idea of the quantity or identity of valid knowledge, except that in a system tending towards entropy (bias, selectivity, inaccuracy) we must assume that the noise will tend to grow and swamp the signal.

Science has appeared to be growing, but the only sure and dependable reality is increasing hype, spin and dishonesty.

The link between scientific stuff and scientific substance has disappeared.

*

In sum: when the signals of science ('publications' or other research communications including spoken words) lose their meaning, when the meaning of science is detached from underlying reality, then there is *no limit to the mismatch.*

Not knowing the truth, the mismatch between truth and un-truth becomes un-measurable – there may be zero correlation between communications and reality.

Scientific communications and underlying reality can be two separate and independent domains. Which explains how real science is collapsing while professional research outputs are booming.

*

The amount of real science (intermittent, infrequent, unpredictable) has surely not stayed absolutely *constant* throughout this inflationary process - but will surely have *declined* due to the environment for real science becoming increasingly hostile.

So, what we call 'science' is an inflating bubble, just skin on gas, and inflating bubbles eventually burst.

And the longer delayed the burst, the bigger the bubble will become (the more gas, the less substance), and the bigger will be the explosion.

*

When the scientific bubble bursts, what will be left over after the explosion?

Maybe only *old science* will prove valid – science from an era when most scientists believed in objective reality and that it was their vocation to discover it; when most scientists were honest and trying to speak the truth – as they understood it - about the natural world.

And from an era when, if scientists had nothing to say for a few years then they said nothing for a few years; when, indeed, if scientists discovered little or nothing then these scientists would state candidly that they had discovered little or nothing.

How far would we need to go back? About two generations or fifty years I should say – to before the mid nineteen sixties, and maybe more...

*

But, in an era of micro-specialization, will there be anyone who can even understand old/ real science, leave aside the problem of finding anyone who can actually *do* real science in the old way?

Scientific validity is about coherence not testing

The problem with science is a problem of validity. Real science had robust (although not infallible) ways of establishing validity; modern professional research cannot establish validity, because it does not recognise any transcendent reality beyond the opinion of 'scientists'.

To be more exact, modern professional research has *methods* which are regarded as intrinsically providing validation – but the methods are themselves unvalidated – indeed the methods used to assert validity are no more than arbitrary conventions enforced by power.

*

Until recently, I usually described real science as being mostly a matter of devising theories which had implications, and testing these implications by observation or experiment.

In other words, science was supposedly about making and testing predictions, devising theories and doing observations and experiments to test them.

This 'classic' view of science is often described as Popperian, being (broadly) based on the work of philosopher Karl Popper – for example in the title of one of his books: *Conjectures and refutations*.

*

Of course there is more which needs to be said to give a sufficient account of Popper's ideas: the predictions must derive from theory, and predictions should be sufficiently complex or non-obvious or counter-intuitive, so as to be unlikely to happen by random chance. And so on.

But it is now clear that this sequence doesn't happen much nowadays, if it ever did.

And, indeed, that there are serious weaknesses about the conceptualization of science as mostly a matter of testing predictions – since this process turns out to be circular – once validation is merely a matter of peer review, of consensus.

*

The main problem is that when science becomes *big*, as it is now, the social processes of science come to *control all aspects of science*, including defining *what counts as a test* of a prediction and *what counts as passing that test*.

Testing now boils-down to the social processes of science, merely.

This means *peer review* = a poll of opinions = government by committees (some actual committees, meeting in a room; some merely virtual committees with participants distributed across time and space). Therefore the validation process is made consensual, and disconnected from any notion of testing putative knowledge in relation to reality.

*

Yet, the problem is therefore not so much at the level of testing – but is at root a problem of coherence,

There can be no 'testing' without coherence. Without coherence there is, indeed, *nothing to test*.

*

Incoherent theories do not have tightly-defined implications and cannot make precise predictions, therefore there is no conceivable way in which they could be put to a test.

So... incoherent theories cannot be tested, and *most* theories in modern research are incoherent (in so far as they are even articulated – there are branches of science operating under the delusion that they do not have any theory), and modern careerist pseudo-science is powerfully resistant to any attempt to create coherence.

When theories are incoherent, hence un-testable, therefore false science *can never be refuted*. The process of (supposedly) 'testing' is one that never ends; nothing can ever be put to a conclusive test, therefore nothing is ever conclusively refuted.

Incoherent 'science' is *not even false*. Therefore incoherent science can be kept going *ad infinitum* – whether it is true or not. From a careerist perspective, therefore, the incoherence of science may be a feature, not a bug.

Science as a sub-species of philosophy

In sum science is a child of philosophy, and as in philosophy, the basic 'test' of science is coherence.

Statements in science ought to *cohere with other statements in science*, and this ought to be checked.

Testing 'predictions' by observation and experiment is *merely one type of checking for coherence*. 'Testing' is, in fact, merely checking for coherence between the predictions of a coherent theory and observations.

This process need not involve a temporal sequence, there no need for prediction to precede the observation testing that prediction, since 'predictions' are (properly) not to do with time but with logic.

Testing in science ought not, therefore, to focus on predictions such as 'I predict now that x will happen under y circumstances in the future' – but instead the focus should be – much more simply – on checking that the various statements of science cohere in a logical fashion.

*

To put it another way: It is an axiom that all true scientific statements are consistent with all other true scientific statements.

True statements should not contradict one another, they should cohere.

In order that coherence not be vacuous, statements must be sufficiently precise in their implications (implications being another word for 'predictions').

So that when it is discovered that there is no logical coherence between two scientific propositions (two theories, 'facts' or whatever), and assuming the reasoning process is sound, *then* one or both propositions must be wrong.

*

Real scientific work is the process of making and learning about propositions.

A newly made proposition that is not coherent with a bunch of previously existing propositions may nonetheless be true, because all or some of the previously existing propositions may be false.

Indeed that is one meaning of a scientific revolution – a revolution is what happens when a new proposition succeeds in overturning a bunch of old coherent propositions, and establishing a new network of coherent propositions: a different set of propositions, coherent on a different basis.

This is always a work in progress, and at any moment there is considerable incoherence in science which is being sorted-out – or, at least, that is the usual assumption.

*

The fatal flaw in modern science is that there is no such sorting-out.

Incoherence is ignored, propositions are merely *piled loosely together* and the result is called a theory.

118

Or the revelation of incoherence is eluded, rather than sorted-out, by the process of micro-specialization and the creation of isolated little worlds *within-which* there may be coherence, but *between-which* there is zero coherence (and no attempt to check or impose coherence).

No such thing as 'Science' anymore

Using this very basic requirement of coherence, it is obvious that much of modern research is not research because it is incoherent – its theories do not make sense, or are obviously wrong.

And furthermore there no coherence between the specialties of research – specialties are not checked against each other. Indeed such checks between specialties are often regarded as impossible – on the basis that different scientific specialties are seen as *incommensurable* (i.e. not measurable against a common standard).

It is not that the propositions of modern professional research are checked and fail the checks, but that no *attempt* whatsoever is made to check for coherence between specialities. Insofar as any need for coherence between micro-specialisms is acknowledged, the actual business of checking is endlessly *deferred...*

Indeed, some philosophies of science have evolved to rationalize the endless deferral of checking for cohesion between specialisms; and there is a big literature in the philosophy of science which purports to prove that different types of science are incommensurable, incomparable, and independent – hence cannot meaningfully be checked against one another.

This implies that there is no unit of scientific validity greater than the micro-specialty. This implies that each micro-specialty (with its narrow selection of foundational assumptions and methods stands alone. This implies that there is *no such thing as 'Science'* and that the individual scientific specialty is the largest possible unit of coherence.

*

If this is true, and it is true in the sense that nobody has even *tried* to demonstrate coherence across and between the 'scientific' research specialities - then *science as a whole does not add-up*.

In other words, there are only the hundreds of microspecialist 'sciences' that cast no light on one another, are irrelevant to each other, do not constrain each other.

This means in turn that all the different micro-specialties that now constitute 'science' would not be contributing to anything greater than themselves considered individually.

This means that, formally speaking, there is *no such thing as Science* only hundreds of 'sciences'– 'Science' is merely an arbitrary collection, a *loose heap* of micro-specialties each yielding autonomous micro-knowledge of unknowable applicability, and the whole given the honorific title of 'Science'.

*

This is very obviously true of modern medical science and biology. For example the massive specialism of 'neuroscience' does not add-up to anything like 'understanding' of the brain or nervous system – it is merely a collection of hundreds of autonomous micro-specialties and factoids about nervous tissue. Observations of this, then of that, then of something else – pile 'em high and call it neuroscience!

These micro-specialties were *not* checked for consistency with each other at any point, and as a consequence they are *not* consistent with each other. Neuroscience was *not* conducted with an aim of creating a coherent body of knowledge, and as a result it is *not* a coherent body of knowledge.

121

'Neuroscience', as a term (it is not a concept, does not rise anywhere *close* to being a concept) is merely an excuse for funding a vast heap of mutual irrelevance.

*

It is not a matter of whether the micro-specialties in modern science are *correct* observations (in the past they probably were honest, nowadays they are quite likely to be dishonest). But that isolated observations – even if honest - are worthless.

Isolated specialties composed of isolated observation are worthless.

Raking-together a heap of worthless observations makes – a worthless heap of observations.

*

It is only when observations and specialties are linked with others (using theories) that consistency can even potentially be checked, whether or not it actually is checked; only then that understanding *might* arise - and then 'predictions' can potentially emerge.

Checking science for its coherence includes testing predictions, and maximizes both the usefulness and testability of science; but a science based purely on testing predictions (and ignoring coherence) will become both incoherent and trivial.

Real science is first coherent, then its coherence is deliberately checked – sometimes (not always) by testing. But modern research is incoherent, and therefore whatever masquerades as checking and testing is not merely irrelevant but actively misleading – merely an excuse for unendingly funding permanently inconclusive research.

Doing science because science is *fun?*

Committed scientists in recent decades have often justified themselves in the face of increasing careerism, fragmentation, incoherence and dishonesty by emphasizing that doing-science (being 'a scientist') is enormous *fun* – and that this is their main motivation for doing it.

*

Although understandable, this is a foolish and indeed *desperate* line of defence. Many things are 'fun' for the people who happen to like them, but fun or not-fun, science was *supposed* to be about reality.

And Hitler, Stalin and Mao seemingly enjoyed being dictators, and redefining 'truth' for their own purposes by the exercise of their power to do so. Perhaps they found all this 'fun' – but does that justify them? Maybe torturers find *their* work fun?

Crosswords, reading romantic novels, getting-drunk, chatting with friends – all these may be *fun*, may indeed be a lot more fun (or, at least, easier fun) than science; but does that justify making them into lifelong careers and spending trillions of dollars on their support and subsidy?

That it may be fun does not justify science.

*

Plus of course science is *not* fun anymore: because being a minor bureaucrat and filling-in forms with lies is not fun (or if it is, the fun is not science); planning your work in detail for the next three years then rigidly sticking to the plan is not fun; being forbidden to do what interests you but forced to do what is funded is not fun; spending your time discussing grants instead of ideas is not fun...

Real science done for vocational reasons *is* (or can be) fun (more exactly, it is profoundly satisfying); but pursuing a modern research career is not science and is not fun.

A modern research career may be rewarding in terms of money, power, status, lifestyle and the like, or sustained by the hope of these – but is not something done for its intrinsic fun-ness.

*

Of course the 'science is fun' line of argument is mostly trying to avoid the 'science is useful' trap.

The usefulness trap must be avoided because the application of science is something intrinsically unknowable. Science is about discovering reality – and knowing this may or may not be useful, may be beneficial or it might well turn out to be *harmful* – indeed fatal; so usefulness cannot be guaranteed.

*

At least usefulness cannot be guaranteed *if you are being honest* – although modern researchers seldom are honest, hence they often do claim that science is predictable, useful and intrinsically beneficial.

(Indeed, in the UK, all government and government-tainted sources of funding require that a successful applicant must make the case that their research is indeed useful and intrinsically beneficial. In other words, the applicant for these sources of money *must lie* in order to be successful. All recipients of such resources are therefore demonstrable liars.)

Modern researchers also sometimes pretend that their kind of science is 'fun' – yet what they are doing is not science, and what they are getting 'fun' from is other stuff entirely: such as the business of trying to get famous, powerful, rich – enjoying the lifestyle of conferences, gossip and intrigue...

*

So real vocational science is 'fun' in the sense of personally rewarding, but this does not justify real science; and almost all of what currently gets called science is neither real nor fun.

After science

The classic science fiction novel *A canticle for Liebowitz* by Walter M Miller portrays a post-nuclear-holocaust world in which the tradition of scientific practice – previously handed-down from one generation of scientists to the next – has been broken. Only a few scientific artefacts remain, such as fragments of electronic equipment.

It turns out that 'after science', scientific objects and records make no sense and are wildly misinterpreted. A blueprint is regarded as if it was an illuminated manuscript, diodes are regarded as lucky talismans.

Modern 'science' has entered a similar state in which the artefacts of science remain – such as places called universities, the academic hierarchy, white coats, laboratory organization, expensive tools and machines, statistical methods, and the peer review mechanism – but understanding of what these *mean* has been lost.

*

A theme associated with philosophers such as the Michaels Polanyi and Oakeshott is that explicit knowledge – such as is found in textbooks and scientific articles – is only a selective summary that omits that the *most important* capability derives from implicit, traditional or 'tacit' knowledge.

It is this un-articulated knowledge – embedded in traditions, habits, practices - that leads to genuine human understanding of the natural world, accurate prediction and the capacity to make effective interventions.

Tacit knowledge is handed-on between- and across-generations by slow, assimilative processes which require extended, relatively unstructured and only semi-purposive human contact.

What is being transmitted and inculcated is an over-arching purpose, a style of thought, a learned but then spontaneous framing of reality, a sense of how problems should be tackled, and a gut-feeling for evaluating the work or oneself, as well as others.

This kind of process was in the past achieved by such means as familial vocations, prolonged apprenticeship, co-residence, and extended time spent in association with a Master – and by the fact that the Master and apprentice personally selected each other.

The Master-apprentice pattern was seen in all areas of life where independence, skill and depth of knowledge were expected: crafts, arts, music, scholarship – and science.

*

Although such methods sound a bit mysterious, not to say obscurationist, to modern ears – in fact they are solid realism and common sense.

Such methods for ensuring the transmission of subtle knowledge recognize the gulf between on the one hand being able to *do* something, and on the other hand knowing *how* you have done it; and the further gap between knowing how you have done something, and being able to teach it by explicit and free-standing *instructions*.

Such systems as apprenticeship recognize that the most important aspects of knowledge may be those which are neither known nor understood to be the most important, or may even be in opposition to that which is believed or supposed to be important.

127

Many things can (tacitly) be *learned* that cannot (explicitly) be *taught*.

*

The educational 'method' was that an apprentice should spend a lot of time with the Master in many situations; and as for educational evaluation, the best way for a Master to know that his skill really has been passed-on, is for him to spend a lot of time with the apprentice in many situations.

Imperfect as it inevitably was, inefficient as it seems; nonetheless traditions *were* as a matter of observable fact maintained and often improved over centuries by means of apprenticeship – which was regarded as the safest and surest way of ensuring that the knowledge and skills could be sustained and developed.

However, modern priorities are different. The preservation and development of high-level human skills and expertise is no longer regarded as a priority, something to which many other concerns will inevitably need to be subordinated.

And the 'Master–apprentice' model of education - which *works*, and which stretches back in human history as far as we know - has been all-but discarded from science (and much of mainstream culture) over recent decades.

*

Indeed the assumptions have now been reversed.

The discarding of traditions of apprenticeship and prolonged human contact in science was not due to any new discovery that apprenticeship was – after all – unnecessary, let alone that the new bureaucratic systems of free-standing explicit aims and objectives, instructions, summaries and lists of core knowledge and

128

competencies, tick-boxes and numerical rating etc. were *superior* to apprenticeship.

Yet there is nothing to suggest that these bureaucratic processes are remotely the equal of apprenticeship: indeed there is nothing to suggest that they work *at all*.

Rather, the Master–apprentice system has been discarded *despite* the overwhelming evidence of its superiority; and has been replaced by the growth of bureaucratic regulation.

*

The main reason for discarding apprenticeship is probably that scientific manpower, personnel or 'human resources' (as they are now termed) have expanded vastly and quickly over the past 60 years – probably about ten-fold.

There was, indeed, zero possibility of such rapid and sustained quantitative expansion being achieved using the labour-intensive apprenticeship methods of the past.

The tradition of apprenticeship was therefore discarded because it stood in the path of the expansion of scientific manpower. Stood, that is to say, in the path of an expansion of *the power of leading scientists* – who in this process evolved from being real scientists into professional research managers.

The choice was between either maintaining the ethos and skills of real science, or else going ahead with the rapid and large scale expansion of research manpower.

The choice that was made was to discard the ethos and skills of science.

*

It has now become implicitly accepted among the mass of professional 'scientists' that the decisions which matter most in science are those imposed upon science by *outside forces*: by employers (deciding who gets the jobs, who gets the promotion), funders (deciding who gets the big money), editors and publishers (deciding who gets their work published in the big journals), bureaucratic regulators (deciding who gets allowed to do what work), and the law courts (deciding whose ideas get backed-up, or criminalized, by the courts) – and of course politicians (deciding the framework within which all these others operate).

It is these bureaucratic mechanisms that constitute 'real life' and the 'bottom line' for modern research practice.

The tradition has been broken.

*

We are living *After Science*, in the same fashion and for similar reasons that philosopher Alasdair MacIntyre recognized (in his 1981 book of that title) that we are living *After Virtue*.

Modern science is in that post-holocaust situation described in *A Canticle for Liebowitz* – the transmission of tacit knowledge has been broken; we have a simulacrum of science but not the reality.

The destruction of real science was concealed by the escalation of hype and spin, thus science was gradually rebuilt as a Potemkin Village: a façade of superficially-impressive pseudo-knowledge concealing a morass of corrupt bureaucracy and mediocre careerism.

Real science in one sentence

*If you are truthful, and spontaneously motivated to spend a lot of time and effort thinking-about and investigating **some thing**, then there is a reasonable chance, but no guarantee, that you will discover **something**.*

Origins of this book

The origins of this book lie in my childhood idealisation of science and its continuation as an ideal well into middle age – and the inexorable dismantling of this ideal by accumulated experience.

What kind of experience?

At first it was noticing that scientists were seldom doing the best work of which they were capable, and becoming aware of their reluctance to take risks or be long-termist in pursuit of scientifically-ambitious work (their preference for high and reliable outputs of mediocre and unimportant work over the smaller chance of major work).

That was dismaying. But primarily it was the experience of non-honesty (that is, *indifference* to honesty) that did for me; the observation that non-honesty was rewarded in career terms.

And finally encountering the systematic imposition of non-honesty; not only indirectly by *rewarding* hype, spin, fashionable incompetence and lies; but eventually directly by the punishment of honesty.

*

Modern scientists are not merely expected to be routinely dishonest when this is expedient (e.g. for career reasons, for the convenience of professional colleagues, for personal and institutional funding etc.); but scientists are now actually forbidden to be truthful all the time and about everything.

Of course, the process is concealed by words, is reframed in apparently more acceptable ways – but I invite anyone who doubts what I say actively to *try* actively practicing science with *scrupulous honesty* - honesty, that is, in *all matters scientific*: even in applications for grants, jobs, promotions, tenure, the presentation of research 'plans', research assessment exercises, press releases and so on. Honest in everything.

Any such individual will almost certainly be confronted with serious problems and intense pressures within one month, probably much sooner.

*

This book is based on one person's knowledge and experience, and is thereby limited in many ways. For example, I do not travel much and have not been to many conferences.

On the other hand, I have worked across an unusually wide range of bioscience and medicine; and for seven and a half years I solo-edited (i.e. no peer review) a large, monthly journal of ideas - international in scope and with very broad-based bioscience content.

I have also known several outstanding real scientists (in the physical sciences, as well as the biosciences and medicine), and some experts in the conduct of real science.

Still, whatever may be their limitations, my knowledge and experience are at any rate *wide enough and deep enough* to draw general conclusions; especially when corruption in science is so very common and so very obvious.

You don't need to be much of a marksman to hit a barn door at five paces...

Further reading and references

The fact that I have not referenced the text of this book comes partly from idleness, partly from the desire to make the reading experience more enjoyable; but mostly from my intention – or at least hope – of *opening eyes to the obvious*, of clarifying the already-known - rather than persuading by weight of (supposed) facts.

(If you *need* persuading, then you *cannot* be persuaded.)

Evidence of the corruption of science, its endemic dishonesty, is all around us and everywhere we look – we need merely to allow the scales to fall from our eyes, need merely to remove our blinkers.

To pick-up and examine *specific items* of dishonesty is merely to diminish the impact of the overwhelming whole by arbitrary, piecemeal and detached consideration.

*

Nonetheless in the past I have tried to do exactly this – to document the corruptions of modern 'science' with referenced papers.

So, for those who want 'evidence', here is a list of my previous publications on themes covered by this book, some of them statistical and historical, replete with a wide range of references to further literature.

All these publications have myself as single author except where otherwise indicated:

- The cancer of bureaucracy: How it will destroy science, medicine, education; and eventually everything else. *Medical Hypotheses*. 2010; 74: 961-965.

- After science: Has the tradition been broken? *Medical Hypotheses*. 2010; 74: 623-625.

- Hunter RS, Oswald AJ, Charlton BG. The elite brain drain. *Economic Journal*. 2009; 119: F231-F251.

- Why are modern scientists so dull? How science selects for perseverance and sociability at the expense of intelligence and creativity. *Medical Hypotheses*. 2009; 72: 237-43.

- The vital role of transcendental truth in science. *Medical Hypotheses*. 2009; 72: 373-376.

- Sex ratios in the most-selective elite US undergraduate colleges and universities. *Medical Hypotheses*. 2009; 73: 127-129.

- Are you an honest scientist? Truthfulness in science should be an iron law, not a vague aspiration. *Medical Hypotheses*. 2009; 73: 633-635.

- Are you an honest academic? Eight questions about truth. *Oxford Magazine*. 2009; 287: 8-10.

- The zombie science of evidence-based medicine: a personal retrospective. *Journal of Evaluation in Clinical Practice*. 2009; 15: 930-934.

- Clever sillies: Why high IQ people tend to be deficient in common sense. *Medical Hypotheses*. 2009; 73: 867-870.

- Pioneering studies of IQ by G.H. Thomson and J.F. Duff - An example of established knowledge subsequently 'hidden in plain sight' *Medical Hypotheses* 2008; 71: 625-628.

- Figureheads, ghost-writers and pseudonymous quant bloggers: The recent evolution of authorship in science publishing. *Medical Hypotheses* 2008; 71: 475-480.

- Zombie science: A sinister consequence of evaluating scientific theories purely on the basis of enlightened self-interest. *Medical Hypotheses* 2008; 71: 327-329.

- First a hero of science and now a martyr to science: The James Watson Affair - Political correctness crushes free scientific communication. *Medical Hypotheses* 2008; 70: 1077-1080.

- Charlton BG, Andras P. 'Down-shifting' among top UK scientists? - The decline of 'revolutionary science' and the rise of 'normal science' in the UK compared with the USA. *Medical Hypotheses* 2008; 70: 465-472.

- Crick's gossip test and Watson's boredom principle: A pseudo-mathematical analysis of effort in scientific research *Medical Hypotheses* 2008: 70: 1-3.

- Charlton, BG, & Andras P. Evaluating universities using simple scientometric research-output metrics: Total citation counts per university for a retrospective seven-year rolling sample. *Science and Public Policy*. 2007; 34: 555-563.

- Peer usage versus peer review. *BMJ*. 2007; 335: 451.

- Measuring revolutionary biomedical science 1992-2006 using Nobel prizes, Lasker (clinical medicine) awards and Gairdner awards (NLG metric). *Medical Hypotheses*. 2007; 69:1-5.

- Which are the best nations and institutions for revolutionary science 1987-2006? analysis using a combined metric of Nobel prizes, Fields medals, Lasker awards and Turing awards (NFLT metric). *Medical Hypotheses*. 2007; 68: 1191-1194.

- Scientometric identification of elite 'revolutionary science' research institutions by analysis of trends in Nobel prizes 1947-2006. *Medical Hypotheses*. 2007; 68: 931-934.

- Boom or bubble? Is medical research thriving or about to crash? *Medical Hypotheses*. 2006; 66: 1-2.

- Charlton B, Andras P. Oxford University's Research Output in the UK context – Thirty-year analysis of publications and citations. *Oxford Magazine* 2006; 254: 19-20.

- Charlton B, Andras P. Oxbridge versus the 'Ivy League': 30 year citation trends. *Oxford Magazine* 2006: 255: 16-17.

- Charlton B, Andras P. Best in the arts, catching-up in science – what is the best future for Oxford? *Oxford Magazine* 2006; 256: 25-6.

- Charlton BG, Andras P. The future of 'pure' medical science: the need for a new specialist professional research system. *Medical Hypotheses*. 2005; 65: 419-25.

- Charlton BG, Andras P. Medical research funding may have over-expanded and be due for collapse. *Quarterly Journal of Medicine*. 2005; 98: 53-5.

Selected sources and acknowledgments

Bronowski, Jacob (1908-1974)

Jacob Bronowski came to my attention when I was fourteen years old, with his stunningly brilliant and heartfelt 1973 television documentary series *The Ascent of Man*. This built-upon a vague childhood interest in science to 'convert' me to 'humanism' – the religion of science as the engine and essence of human progress. Afterwards I read and re-read pretty much everything Bronowski ever wrote, but especially *The Common Sense of Science* (1951) and *Science and Human Values* (1956). I would now regard Bronowski as an (intrinsically un-replicable) *transitional* figure between science based in orthodox religiousness and careerist research derived from modern nihilism. Bronowski was raised a Jew and later became a militant atheist – but he always retained his inculcated religious devotion to transcendental values (especially truth but also beauty). Despite later divergences, I got from Bronowski the critical insight that science depends utterly on 'the habit of truth' – that truthfulness is non-optional, an iron law.

Calne, Roy Yorke

From conversation and reading his autobiography *The Ultimate Gift* (1998), the great scientist and heroic surgical pioneer Roy Calne made me recognize how rapidly and radically medical research had changed and become corrupted during the decades of his career.

Crick, Francis (1916-2004)

Although I was aware of him from my mid-teens, it was after 1994 when I embarked on a career change as a theoretical scientist, that I regarded Francis Crick as a spiritual mentor and model. I was particularly influenced by his autobiography *What mad pursuit* (1988).

Gregory Clark

Economic historian, University of California at Davis. Very few books have had such impact on my thinking as *A farewell to alms: a brief economic history of the world*. I didn't really believe that there was anyone alive capable of such large scale thinking; and I later had the good fortune to meet Clark and have several in-depth discussions.

Einstein, Albert (1879-1955)

Naturally, Einstein always had for me god-like status as the pinnacle of modern scientific achievement and living – and I have read innumerable biographies and memoirs – however it all began with an autobiographical essay in a multi-author volume called *I believe: the personal philosophies of twenty-three eminent men and women of our time* (1940) that I found on our bookshelves at home, in a wartime edition.

Feynman, Richard (1918-1988)

Feynman was a later influence on me – probably from around 1988; as a representative of uncompromising, fearless honesty and scientific integrity yet without (apparently) any trace of transcendental belief. Like Einstein, Feynman was raised as a Jew but became very much a modern 'liberated' man. Yet, it is not possible consistently to use Feynman as a mentor, since his personal decision to be utterly truthful had no deeper rationale, and has no traction in the corruptly dishonest world of modern research. Feynman got away with truthfulness for a long time, due to his personal charm, terrifying brilliance and Nobel Prize – but I now see him as one of the last of a transitional generation of religious converts to non-nihilistic atheism, after which careerism and full-nihilism took-over completely.

Hannam, James

A popular book by Hannam called *God's philosophers* (2010) – although not a major influence – effectively hammered-home for me the substantial and undeniable achievements of medieval science – that is, of science before there was such a thing as 'science'.

Healy, David

Since I encountered his *Suspended Revolution* (1991) I have avidly been reading the work of David Healy, and his books on the history, sociology and practice of psychiatry, psychopharmacology and medicine generally – including his detailed and conclusive documentation of the thorough corruption of research and practice especially since the mid-1960s. In my estimation, Healy is incomparably the greatest writer on these topics in the English language, and it has been a privilege to know him as a friend and (in a small way) as a collaborator.

Hesse, Herman (1877-1962)

Hesse was author of *The Glass Bead Game* (1943) also known as *Magister Ludi* and *Das Glasperlenspiel*. This is a memorable account of the fascination of formal intellectual activity for its own sake, as an utterly absorbing 'game' – cut-off from the real world, from usefulness, from practical applications.

Horrobin, David L (1939-2003)

David Horrobin founded *Medical Hypotheses* on Popperian principles, and bequeathed the journal's editorship to me. He was one of the last classic scientists to succeed in medical research – and in order to do so he needed to leave the conventional academic structure and self-fund via his business activities in pharmacology (he also become the object of extraordinary resentment). Horrobin was the first to notice and document that the rate of clinical innovation had declined since the mid 1960s - Horrobin, D.F. Scientific medicine - success or failure? In: Weatherall, D.J.; Ledingham, J.G.G.; Warrell, D.A. (Eds.) *Oxford Textbook of Medicine*, 2nd Edn. Oxford University Press, Oxford. 1987: 2.1-2.3.

Hull, David L. (1935-2010)

Hull's great work *Science as a process* (1988) ingeniously uses the theory of evolution as its example of how science can be conceptualized as a process of evolution by natural selection – with science regarded in terms of the replication of theories and professional status as the main evaluation. Hull's empirically-dense account seems true of classic or 'real' science – and he therefore assumed that scientific status was *constrained* by reality. However since Hull's explanatory model has no reference to transcendental truth (real reality) as a regulatory ideal; the evolutionary concept of science turns-out to be equally true of generic professional research of all types – including 'zombie' science. I knew Hull somewhat, having met and corresponded with him – and he was on the *Medical Hypotheses* editorial board; and my impression was that in late life he recognized with considerable dismay that science had (since the period described in his main book) undergone a 'turn' and evolved away from a concern with the real world and into autonomous careerism.

Michael Mabe, Mayur Amin.

Growth dynamics of scholarly and scientific journals. Scientimetrics. 2001; 51: 147-162.

Oakeshott, Michael (1901-1990)

Rationalism in Politics, 1962.

Polanyi, Michael (1891-1976)

Personal Knowledge, 1958.

Popper, Karl (1902-1994)

I came across the work of Popper in my late teens via the book of that title by Bryan Magee (1973) – I went on to read some of Popper's own work, and particularly liked his autobiography *Unended Quest* (1977 edition). Popper's normative description has been vastly influential in British science, and it is still used to tell undergraduates how to structure their investigations. It was also the major influence on the founding by David L Horrobin of *Medical Hypotheses*, the theoretical journal I edited from 2003-2010 – Popper had indeed been a member of the editorial advisory board from about 1974 until his death. Yet, for all the lip service paid to Popper, it is clear that his ideas have zero real influence on modern mainstream science; probably because Popper, like most other philosophers of science, tried to describe science purely in terms of its process and pragmatic value to the happiness and comfort of society; and without regard to its transcendental aims.

Rees, Jonathan

Professor of Dermatology, University of Edinburgh – previously at Newcastle University. For many years, at varying intervals, I have met and talked with Jonathan about science and medicine in broad and specific terms. Despite considerable professional success, he has - so far as I can tell - remained completely honest.

Romer, Paul

I got the (mistaken) perspective of 'economic growth fuelling science fuelling economic growth' from an audio interview with Romer on econtalk.org dated 27 August 2007. Indeed, I had co-argued something similar myself (with Peter Andras) in a book called *The Modernization Imperative* (2003).

Watson, James D

I read Watson's *The Double Helix* about 1975-ish and have re-read it many times since; and also enjoyed numerous other accounts of the discovery of the structure of DNA – especially the 1987 TV movie *Life Story*. Also influential more recently was the essay: J. Watson, Succeeding in science: some rules of thumb, *Science* 1993; 261: 1812.

Ziman, John (1925-2005)

Ziman's early books such as *Public Knowledge* (1968) and *Reliable Knowledge* (1978) were historical and sociological descriptions of classic science of the golden age; but Ziman really distinguished himself by being (so far as I know) the first thoroughly to document the death of classic science in *Real Science*, published in 2000. Here he described the transformation of 'academic' science into 'post-academic' science. In this book I have re-named Ziman's academic science as 'real' science and his 'post-academic' science as 'professional research'. I knew Ziman slightly when he was on the editorial board of *Medical Hypotheses*, and am pleased to dedicate this book to his memory.

www.ingramcontent.com/pod-product-compliance
Lightning Source LLC
Chambersburg PA
CBHW021931190326
41519CB00009B/982